Biological Waste Treatment

Biological Waste Treatment

W. W. ECKENFELDER, Jr
AND
D. J. O'CONNOR

*Manhattan College
Civil Engineering Department
New York*

PERGAMON PRESS

OXFORD • LONDON • EDINBURGH • NEW YORK
TORONTO • SYDNEY • PARIS • TOKYO • BRAUNSCHWEIG

1961

PERGAMON PRESS LTD.
Headington Hill Hall, Oxford
4 & 5 Fitzroy Square, London W. 1

PERGAMON PRESS (Scotland) LTD.
2 & 3 Teviot Place, Edinburgh 1

PERGAMON PRESS, INC.
44-01 21st Street, Long Island City, New York 11101

PERGAMON OF CANADA LTD.
6 Adelaide Street East, Toronto, Ontario

PERGAMON PRESS (Australia) PTY. LTD.
20-22 Margaret Street, Sydney, New South Wales

PERGAMON PRESS S.A.R.L.
24 rue des Ecoles, Paris 5e

VIEWEG & SOHN G.m.b.H
Burgplatz 1, Braunschweig

Copyright © 1961
Pergamon Press Ltd.

Second Printing, 1964
Third Printing, 1966

Library of Congress Card Number 61—10913
08 009547 X

Printed in the United States of America

CONTENTS

Chapter		Page
	PREFACE	vii
1	WASTE CHARACTERISTICS AND TREATMENT METHODS	1
	Analysis of Data	5
	Waste Treatment Processes	7
	Transfer and Rate Mechanisms	9
	Biochemical Oxygen Demand	10
2	PRINCIPLES OF BIOLOGICAL OXIDATION	14
	BOD Removal and Sludge Growth	16
	Summary of Bio-oxidation Kinetics	28
	Bio-oxidation of Pure Compounds	33
	Character of Biological Sludge	34
	Oxygen Utilization	39
	Sludge Production and Oxidation	53
	Nutritional Requirements	61
	Effect of Temperature	67
	Effect of pH	69
3	THEORY AND PRACTICE OF AERATION	76
	Oxygen Saturation	76
	Theory of Oxygen Transfer	78
	Stream Aeration	82
	Bubble Aeration	86
	Measurement of Oxygen Transfer Coefficients	95
	Diffused Aeration	97
	Mechanical Aeration	112
4	STREAM AND ESTUARY ANALYSIS	122
	Oxygen Demands	123
	Oxygen Resources	128
	Oxygen Balance in Streams	131
	Oxygen Balance in Estuaries	135
5	SOLID–LIQUID SEPARATION	152
	Sedimentation	152
	Zone Settling and Compression	167
	Flotation	179

CONTENTS

6	AEROBIC BIOLOGICAL TREATMENT PROCESSES	188
	Lagoons and Stabilization Basins	188
	Activated Sludge	203
	Trickling Filters	221
7	ANAEROBIC BIOLOGICAL TREATMENT PROCESSES	248
	Theory	248
	Digestion Design	252
	Anaerobic Decomposition of Industrial Wastes	264
8	SLUDGE HANDLING AND DISPOSAL	270
	Sludge Drying Beds	270
	Vacuum Filtration	273
	AUTHOR INDEX	290
	SUBJECT INDEX	293

PREFACE

THIS book is the outgrowth of a course entitled "Bio-oxidation of Organic Wastes—Theory and Design" initiated at Manhattan College in 1955. The objective of the course was to present the fundamentals of bio-oxidation which would serve as a framework for the analysis, design and operation of biological waste treatment facilities.

This book reflects the authors' approach to the solution of waste treatment problems. It is not intended as a scientific treatise on the physics, chemistry and biology of waste treatment but rather as an engineering text applying principles of the aforementioned fields to the design and operation of waste treatment facilities. The authors have employed scientific principles which have been mathematically formulated and empirical relationships where necessary. Although this text is primarily intended to serve as a guide for the practising engineer, it should also serve as a useful reference for graduate students in sanitary engineering.

Due to the rapid expansion of the science and technology of waste treatment it is assumed that many of the concepts described in this book will be further developed and refined by the competent researchers from whose work the authors freely drew.

The authors acknowledge the reproduction of many figures and tables taken from the publications: *Sewage and Industrial Wastes*; *Water and Sewage Works*; *Transactions of the American Society of Civil Engineers* and *Proceedings of the Purdue Industrial Waste Conference*. Reproductions from the manufacturers of waste treatment equipment are acknowledged in the text.

A word of appreciation is due to the secretaries of the Civil Engineering Department, Berenice Maguire and Ruth Minihan, and to the many students who assisted in the preparation of this manuscript.

CHAPTER 1

WASTE CHARACTERISTICS AND TREATMENT METHODS

MANY problems in the pollution of natural waters are attributable to municipal sewage and organic industrial wastes. The characteristics of these wastes which are significant in pollution are the suspended solids, oxygen demand of the organic matter and the coliform bacteria. The steps involved in a pollution abatement program are stream and waste sampling, laboratory and pilot plant investigations, and process design and plant construction. The magnitude of the problem is quantitatively defined by a survey and an analysis of the stream to determine the ability of the receiving water to assimilate the various pollutional elements. Most water control agencies have established minimum stream standards for the various rivers under their jurisdiction. This phase is followed by a survey to determine the quantity and characteristics of the wastes. It involves a gaging, sampling and analysis program. Frequently, a laboratory and pilot study are necessary to determine a feasible method of treatment. These results are used to define process design criteria for full-scale treatment.

The pollutional characteristics of waste waters may be classified according to their state (suspended, colloidal, and dissolved) and their nature (inorganic, organic, gases, and living organisms). The significant characteristics of organic wastes are usually the suspended solids and the organic content. The organic content is measured in terms of its oxygen equivalence by means of the BOD test. The BOD may be defined as that quantity of oxygen required during the stabilization of decomposable organic matter and oxidizable inorganic matter by aerobic biological action. In the case of municipal sewage, the coliform density is also very significant, although this factor is usually not present in most industrial wastes. These characteristics are used as criteria of water quality in the receiving streams. Characteristics of some typical organic wastes are shown in Table 1-1.

TABLE 1-1. CHARACTERISTICS OF SOME TYPICAL ORGANIC WASTES

	Suspended solids, p.p.m.	BOD_5, p.p.m.	pH
Sewage	100–300	100–300	—
Pulp and paper	75–300	—	7·6–9·5
Dairy	525–550	800–1,500	5·3–7·8
Cannery	20–3,500	240–6,000	6·2–7·6
Packing house	650–930	900–2,200	—
Laundry	400–1,000	300–1,000	—
Textile	300–2,000	200–10,000	—
Brewery	245–650	420–1,200	5·5–7·4

The quantity, as well as the characteristics of the wastes, is important in defining the magnitude of the pollution problem. The quantity of several waste waters is shown in Table 1-2.

An industrial waste survey may be required to define the pollutional characteristics. The frequency of sampling is usually established by the nature of the process, i.e. batch, intermittent, or continuous. Individual or short-time composite samples on the most important waste characteristics, i.e. BOD, COD, and suspended solids, should be collected and daily composite samples on all other characteristics. All composite samples should be weighed according to flow. Oxygen demand, solids, flow and nutritional content are necessary factors to determine the waste treatment process design. Characteristics such as pH, temperature, toxic elements, etc., are vital to the design of pretreatment facilities and instrumentation. Simplified analytical procedures are desirable for laboratory analysis and control. For example, the COD test may frequently be substituted for the BOD or volatile solids test provided a significant correlation can be established between the two characteristics.

Waste flow data are frequently available from plant records. In the absence of records or metering devices, it is necessary to install such means which can be accommodated by the physical features of the plant. In batch processes, time and volumetric readings are

TABLE 1-2. QUANTITY OF REPRESENTATIVE PROCESS WATERS

Paper		12,000–15,000 gal/ton pulp paper 25,000–30,000 gal/ton high grade
Tannery		8–10 gal/hide/day
Meat (slaughter house)	cattle hogs	400 gal/animal 150 gal/animal
(packing house)	cattle hogs	2,000 gal/animal 700 gal/animal
Cannery	corn beans squash peaches pears apricots tomatoes beets peas	40 gal/case No. 2 cans 70 gal/case No. 2 cans 20 gal/case No. 2 cans 90 gal/case No. 2 cans 90 gal/case No. 2 cans 80 gal/case No. 2 cans 50 gal/case No. 2 cans 30 gal/case No. 2 cans 25 gal/case No. 2 cans
Milk		100–300 gal/100 lb milk
Brewing		500 gal/bbl.
Textiles	bleaching dyeing	3,000–4,000 gal/100 lb cotton 250–2,000 gal/100 lb cotton
Gasoline	(wash)	60,000 gal/1,000 gal

adequate, whereas in continuous processes, appropriate measuring devices are necessary. In order to derive design criteria, flow measurements should be taken not only on the total waste, but also on the individual sources of waste throughout the plant. The average daily flow is used as the basis for the process design and a maximum flow used for determination of the hydraulic capacity of the facilities. A diagram indicating the sampling points for an industrial waste survey of a cannery is shown in Fig. 1-1.

The data on flow and waste characteristics as collected and measured at the locations indicated in Fig. 1-1 provide the basis for

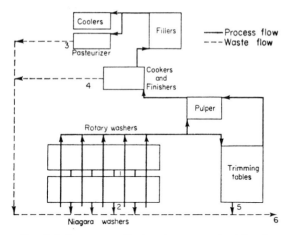

FIG. 1-1. Sewer map of a tomato process plant showing sampling stations.

calculating a flow and material balance. This balance for average conditions for the cannery shown in Fig. 1-1 is:

Waste sources (see FIG. 1-1)	1	2	3	4	5	6
Flow, gal/min	560	560	150	75	200	1,520
COD, lb/day	2,210	3,710	371	9,050	384	15,725
SS, lb/day	870	1,570	27	2,995	78	5,540

An analysis of a flow and material balance frequently indicates an economical solution to the waste treatment problem by eliminating or reducing the quantity of waste matter or waste water. A reduction of this nature can be accomplished by the following methods:

(1) Recovery or utilization of waste products.

(2) Removal of waste matter in a dry state.

(3) Segregation of non-contaminating waste water such as cooling water. Non-contaminated process water can frequently be removed from the waste system for direct discharge to the receiving body. By such segregation, the hydraulic load to the waste treatment plant may be reduced.

(4) Recirculation of waste waters. Some processes do not require a water of high quality and, therefore, could use as supply the waste water from other units in the plant. Recirculation of this nature reduces the hydraulic load and concentrates the waste load.

ANALYSIS OF DATA

The strength and flow characteristics of most wastes are highly variable and these data are usually susceptible to statistical analysis. These variations may be caused by production changes in continuous processes, and by random discharges in batch operations. An example of batch operation, its flow pattern, and the statistical variation of flow data are shown in Fig. 1-2.

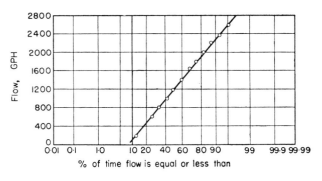

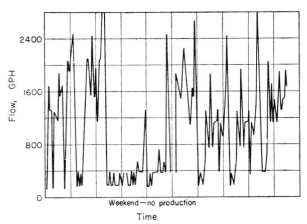

FIG. 1-2. Variation in flow from a batch operation.

Statistical analysis of variable data provides the basis for process design. The probability of occurrence of any value of flow, BOD, suspended solids, etc., may be determined as follows:

SS	BOD	m	$\dfrac{m}{n+1}$	$\%$ occurrence
48	200	1	0·1	10
83	225	2	0·2	20
85	260	3	0·3	30
102	315	4	0·4	40
130	350	5	0·5	50
134	365	6	0·6	60
153	430	7	0·7	70
167	460	8	0·8	80
180	490	9	0·9	90

The suspended solids and BOD values are each arranged in order of increasing magnitude. n equals the total number of solids or BOD values and m is the assigned serial number from 1 to n. $m/(n+1)$ is the plotting position, which is equivalent to the percent occurrence of the value. The actual values are then plotted against the percent occurrences on probability paper as indicated on Fig. 1-3. A line of best fit may usually be drawn by eye. If desirable, it may be calculated by standard statistical procedures. The probability of occurrence of any value can now be obtained: for example, in Fig. 1-3,

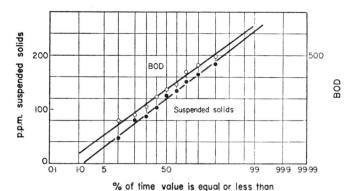

Fig. 1-3. Probability of suspended solids and BOD in a raw waste.

the BOD will be equal to or less than 500 p.p.m. 88 per cent of the time. Flow and waste characteristics are at times preferably correlated to production units. A statistical analysis of the various waste characteristics provides the basis for the assignment of the process and hydraulic flows. Process design criteria are usually based on the 50 per cent values while the 90 per cent and 99 per cent range is used for the hydraulic design.

WASTE TREATMENT PROCESSES

The treatment of organic wastes may usually be classified by three steps, viz. pretreatment, biological oxidation and sludge treatment and disposal. Pretreatment includes screening, grit removal and sedimentation or flotation. Screening is employed in cases where the waste contains a significant quantity of large solids. Domestic sewage and many cannery wastes are typical examples of wastes susceptible to this method of treatment. When the waste contains inorganic suspended solids, a grit chamber is used as a primary separation unit. The removal of sand and grit is advisable in order to avoid excessive wear on mechanical units and to prevent the cementing effect in the sludge blanket of the primary settling tanks. When the waste contains suspended solids, sedimentation and flotation units are employed. These units also serve to reduce the organic load on subsequent biological treatment facilities. Sludge removed from the settling or flotation tanks is further treated in separate units.

Biological oxidation is usually accomplished in fixed-bed units (trickling filters) and in fluid-bed systems (activated sludge). It is employed in removing the colloidal and dissolved organic matter. A trickling filter generally consists of a circular structure containing a bed of suitable depth of crushed stone, crushed slag or other reasonably hard and insoluble medium. Wastes are sprayed over the bed by rotary distributors or fixed nozzles. The liquid trickles downward through the bed to the underdrains. Underdrains are usually precast vitrified clay blocks with suitable openings for drainage and ventilation. As organic wastes are continuously passed over the medium a gelatinous film of micro-organisms develops on the surfaces of the medium. This film is composed of microbial cells in various stages of activity, organic matter in various stages of decomposition, and oxidized and non-oxidizable organic and inorganic residues. Colloids are adsorbed by the film, where they are coagulated and hydrolyzed to smaller molecules, and soluble organics diffuse into the film, where they are oxidized by enzymatic action.

Biological filters may be designated as either standard or high-rate trickling filters. Standard filters operate at a low hydraulic and organic loading and produce a highly stabilized effluent. High-rate filters have relatively high hydraulic and organic loading. Filters are followed by settling units which remove oxidized organic matter and microbial solids which have sloughed off the filter.

The activated sludge process may be defined as a system in which flocculated biological growths are continuously circulated and contacted with organic waste in the presence of oxygen. The oxygen is usually supplied from air bubbles injected into the sludge-liquid mass under turbulent conditions. The process involves an aeration step followed by a solids–liquid separation step from which the separated sludge is recycled back for admixture with the waste. A portion of this sludge is removed for further treatment and disposal.

Sludge from primary and secondary units requires further treatment before disposal. This sludge is frequently too dilute for economical treatment and therefore is concentrated by gravity thickening or flotation. This step is followed by either digestion or dewatering. Digestion, anaerobic or aerobic, reduces the volatile content of sludge. Anaerobic digestion consists of liquefaction and gasification of the organic content and results in end-products of methane, carbon dioxide and stabilized residue. In aerobic digestion, the organic solids are progressively oxidized by a biological mechanism similar to the activated sludge process. Aeration is required to provide oxygen and mixing for the process. The sludge produced from the aerobic process is similar to that of the anaerobic process.

Sludge dewatering may be accomplished by air drying, vacuum filtration or centrifugation. In air drying, the sludge may be dried on open or closed sand beds. Dewatering results from filtration and evaporation. The filtrate is returned to the process and the dried sludge periodically removed for final disposal. In vacuum filtration, the sludge liquid is removed through a porous medium under an applied vacuum. The dried sludge cake is continuously removed from the filter drum. The filtrate is returned to the process. High-speed centrifugation is occasionally used for sludge dewatering.

WASTE CHARACTERISTICS AND TREATMENT METHODS 9

Dewatered or dried sludge is finally disposed of by incineration or land fill.

Depending on the nature of the waste and the degree of treatment required, these unit processes are combined into three general classifications depending upon the efficiency of removal of BOD and suspended solids from domestic sewage:

Class	Percent removal	
	SS	BOD
Primary	50–70	20–40
Intermediate	70–80	40–75
Complete	80–95	75–95

For industrial wastes, these efficiencies will depend upon the relative quantities of suspended and dissolved organic matter.

TRANSFER AND RATE MECHANISMS

In biological treatment, the two general concepts which are used to formulate the rates of biological oxidation and reduction are mass transfer and rate kinetics. Mass transfer is a diffusion process involving the exchange of material from one phase to another. The rate of exchange depends upon a driving force, which is a differential equation:

$$\frac{dq}{dt} = \frac{F}{R} \qquad (1\text{-}1)$$

in which q = quantity transferred; t = time; F = driving force; and R = resistance.

A pertinent example in biological treatment is aeration, which may be defined as the transfer of oxygen from the gas phase to the liquid phase. The driving force is the difference between the interfacial concentration of dissolved oxygen and that in the body of the liquid. The resistance to transfer is located at the interface between the gas and the liquid. Mass transfer is influenced by the physical and chemical characteristics of the two phases, such as temperature, viscosity and surface tension.

Rate kinetics is the study of the velocities of chemical and biochemical reactions. A characteristic equation defining reaction rates is:

$$\frac{dc}{dt} = K f(c) \qquad (1\text{-}2)$$

in which c = concentration of reacting constituent; t = time; $f(c)$ = a function of the concentrations of reacting substances; and K = reaction-rate constant.

The reaction-rate constant depends upon temperature and the characteristics of the solution. Many reactions of interest in biological treatment are usually defined as zero order or first-order reactions. In the former case, the rate is independent of concentration and dc/dt is constant, while in the latter the rate is proportional to concentration of the reacting substance. The oxidation of organic matter at low concentrations is usually represented by a first-order reaction. When considering complex organic wastes, the over-all rate is determined by the individual reaction rates and results in a time-dependent rate coefficient.

BIOCHEMICAL OXYGEN DEMAND

The oxygen demand of a waste is quantitatively evaluated by the biochemical oxygen demand test. The biochemical oxygen demand (BOD) is the amount of oxygen required by the living organisms engaged in the utilization and ultimate destruction or stabilization of the organic matter. This test measures the combined effects of all the putrescible substances of which a particular waste may be composed. The BOD exerted is evidenced by two stages; the first, in which the carbonaceous matter is oxidized and the second, in which the nitrogenous substances are oxidized. Complete stabilization requires such a long period of incubation that it is impractical for routine analysis. A 5-day period of incubation is recommended as a standard procedure and BOD results are usually reported on this basis. In many cases it is necessary to determine the ultimate BOD and this may be accomplished by evaluating the rate at which the reaction proceeds to completion. The rate of oxidation of many unstable chemical substances can be estimated by a first-order reaction. When more than one substance is involved the reaction rate may be second-order. In the case of sewage and industrial rates, it appears that a first-order reaction rate reasonably well defines the oxidation of organic matter in the first stage. A first-order reaction is one which is characterized by a rate directly proportional to the concentration of the substance reacting. Numerous investigators have objected to the use of a first-order reaction as a measure of biochemical activity. The commonly accepted mechanism of oxidation consists of two biochemical reactions—first, the rapid rate of growth of bacterial cells by the assimilation of the organic matter, and second, the relatively slow rate of subsequent oxidation of these cells. The first phase generally is completed in from 12 to 60 hr, depending upon the lag. In some cases it may even be completed before the sample is introduced into the BOD bottle. Up to the present time, no mathematical model has been constructed on a theoretical basis and consequently, most practitioners have continued to use the first-order reaction to define the rate of the BOD reaction. This equation may be expressed as follows:

$$\frac{dL}{dt} = -K_1 L \qquad (1\text{-}3)$$

in which L is the concentration of the substance reacting and K_1 the reaction constant. The equation may be integrated to the following results:

$$\left. \begin{array}{l} \log_e \dfrac{L_t}{L_0} = K_1 t \\[1em] \log_{10} \dfrac{L_t}{L_0} = k_1 t \end{array} \right\} \qquad (1\text{-}4)$$

in which L_0 = initial concentration of organic matter or ultimate oxygen demand; L_t = concentration of organic matter remaining at the end of time, t; K_1 = reaction coefficient base e; k_1 = reaction coefficient to base 10; and $Y = L_0 - L_t$.

It is to be noted that the above equations refer to organic matter measured in terms of oxygen remaining at the end of any time period.

It follows that the organic matter oxidized, or the oxygen used, is equal to:

$$Y = L_0 - L_t \qquad (1\text{-}5)$$

Substitution of Equation (1-5) in Equation (1-4) and rearranging terms, there results:

$$Y = L_0(1 - 10^{-k_1 t}) \qquad (1\text{-}6)$$

in which Y is the BOD exerted in time, t.

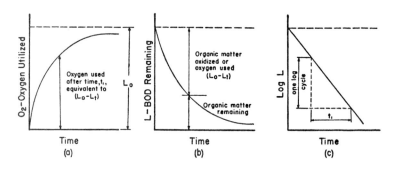

Fig. 1-4. General BOD curves.

These equations are shown graphically in Figs. 1-4 and 1-5. The temperature influence on this reaction is reflected on the coefficient k_1 as follows:

$$k_t = k_{20} \times 1\cdot047^{T-20} \qquad (1\text{-}7)$$

A common range of values of k_1 is 0·10 to 0·30 per day for municipal sewage and many industrial wastes. A number of methods are available for the determination of the coefficient k_1 from laboratory data. Knowledge of this rate permits determination of the ultimate demand from the 5-day value, commonly reported.

In conducting the laboratory tests for the BOD, particular attention should be given to the proper seeding material. "Standard Methods" recommends sewage seed, but the disadvantage of employing this

WASTE CHARACTERISTICS AND TREATMENT METHODS 13

seed for certain industrial wastes is recognized. A more correct estimation of the BOD may be obtained if water from the receiving body downstream from the point of discharge is used. The use of

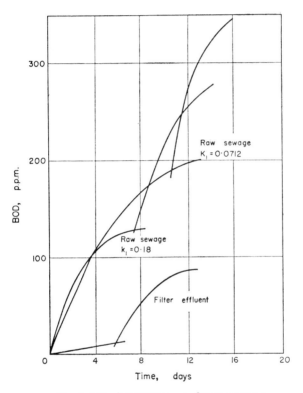

FIG. 1-5. Typical BOD curves for raw sewage and a biological filter effluent.

this seed may stimulate the nitrification stage. River samples are thus automatically seeded. If, however, sewage seed is used, the 5-day values may differ, although the ultimate determined by either seeding method should be approximately equal in magnitude.

CHAPTER 2

PRINCIPLES OF BIOLOGICAL OXIDATION

WHEN an organic waste is contacted with biological sludge, BOD is removed by several mechanisms. Suspended and finely divided solids are removed by adsorption and coagulation. A portion of the soluble organic matter is initially removed by absorption and stored in the cell as a reserve food source. Additional dissolved organic matter is progressively removed during the aeration process, resulting in the synthesis of sludge and the production of carbon dioxide and water. Availability for oxidation decreases as the complexity of the organic compounds increases. Large particles undergo subdivision by hydrolysis prior to oxidation. The rate of BOD removal after initial absorption depends primarily upon the concentration of BOD to be removed and the concentration of sludge solids. The reactions involved in the removal of BOD from solution during bio-oxidation can be interpreted as a 3-phase process (Weston and Eckenfelder, 1955):

(1) An initial removal of BOD on the contact of a waste with a biologically active sludge which is stored in the cell as a reserve food source.

(2) Removal of BOD in direct proportion to biological sludge growth.

(3) Oxidation of biological cellular material through endogenous respiration.

These reactions are illustrated by the following equations:

Organic Matter Oxidation

$$C_xH_yO_z + O_2 \xrightarrow{enzyme} CO_2 + H_2O - \Delta H \qquad (2\text{-}1)$$

Cell Material Synthesis

$$(C_xH_yO_z) + NH_3 + O_2 \xrightarrow{enzyme} \text{cells} + CO_2 + H_2O - \Delta H \qquad (2\text{-}2)$$

PRINCIPLES OF BIOLOGICAL OXIDATION

Cell Material Oxidation

$$(\text{cells}) + O_2 \xrightarrow{\text{enzyme}} CO_2 + H_2O \quad (2\text{-}3)$$
$$+ NH_3 - \Delta H$$

The term ΔH represents the heat of reaction. These generalized equations must be modified for organic compounds containing nitrogen or sulfur.

Equation (2-1) is the conventional equation of combustion. If nitrogen is present, it will be oxidized to nitrate; sulfur will be oxidized to sulfate.

Equation (2-2) represents the synthesis of cell material from organic substrates.

Equation (2-3) represents the oxidation of cellular material previously synthesized.

The synthesis of activated sludge as shown by Equation (2-2) employs ammonia as a source of nitrogen. Equations (2-1) to (2-3) can be illustrated using lactose sugar as a source of carbon (Hoover and Porges, 1952). The complete oxidation of lactose is:

$$C_{12}H_{22}O_{11} \cdot H_2O + 12\,O_2 = 12\,CO_2 + 12\,H_2O$$

or

$$(CH_2O) + O_2 = CO_2 + H_2O$$

The synthesis reaction (Equation 2-2) may be illustrated using 8 sugar units.

$$8(CH_2O) + 3\,O_2 + NH_3 = C_5H_7NO_2 + 3\,CO_2 + 6\,H_2O$$

The cells produced will undergo oxidation (Equation 2-3)

$$C_5H_7NO_2 + 5\,O_2 = 5\,CO_2 + NH_3 + 2H_2O$$

The validity of these equations was confirmed by manometric observations.

Tamiya (1935) has stated that all cell synthesis reactions are exothermic and hence energy is supplied by the reaction. Exact quantitative relations can be determined only by experiment since they will vary depending upon the specific environment.

BOD REMOVAL AND SLUDGE GROWTH

The growth of a biological sludge mass in a batch oxidation follows the sigmoidal curve shown in Fig. 2-1. This type of growth is followed by all biological populations. The lower portion of the growth curve is concave upward and represents a geometric increase in sludge mass (a-b). This is called the logarithmic growth phase, during which regular and maximum multiplication of the sludge cells is taking

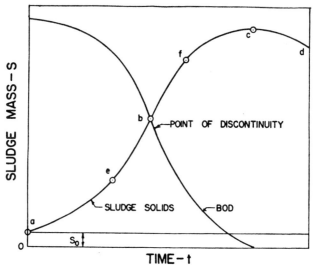

FIG. 2-1. BOD removal and sludge growth relationships.

place. This growth phase occurs in the presence of an abundant supply of food. The middle portion of the curve is approximately linear. As the available food supply becomes exhausted, a declining growth phase occurs in which cellular division occurs at less frequent intervals (b-c). The upper portion of the curve (b-c) follows a first order reaction. The sludge growth curve becomes asymptotic to a limit which is dependent upon the concentration of available food.

The portion of the growth curve (c-d), following the sigmoidal curve, represents the decrease in sludge mass resulting from auto-oxidation which occurs after the depletion of the available food. This is often called the endogenous respiration phase of activated sludge.

The auto-oxidation initially follows first-order kinetics, followed by a decreasing oxidation rate as the bacterial substrate becomes less available for oxidation.

Lag Phase

In some cases a lag phase may exist in the growth relationship. This will occur when a dissimilar food supply is introduced (i.e., sewage organisms to industrial wastes) or by employing sludge which is in the advanced endogenous phase. A lag phase of 2–5 days to attain full purification capacity was found by Sawyer *et al.* (1955) when employing a sludge aerated under starvation conditions for 21–22 days.

Initial Removal

In many biological systems a very high rate of removal of BOD is observed immediately after contact of waste with sludge. Suspended and colloidal solids are removed by flocculation and adsorption while soluble organic matter is removed by biosorption. The magnitude of this high rate depends on the nature of the waste and the characteristics of the sludge. In some cases, this high initial rate exceeds the maximum rate of sludge growth and BOD is stored in the microbial cell. This stored BOD is subsequently oxidized over several hours aeration of the sludge.

It is frequently possible to formulate the initial removal by considering the high initial removal over a short time-interval (10–15 min). In this case, only that portion of the BOD which is removable by the biological sludge over the specified time-interval is considered. Eckenfelder (1959) employed the relationship:

$$-\frac{dL}{dS} = K_i L \quad (2\text{-}4a)$$

which integrates to:

$$\frac{L_{ri}}{L_t} = 1 - e^{-K_i S} \quad (2\text{-}4b)$$

in which K_i = initial removal rate coefficient; S = initial biological sludge solids; L_{ri} = BOD removed over specified time interval; and L_t = maximum BOD removable over specified time interval.

Initial BOD removal for a pulp and paper mill waste and for a mixture of domestic sewage and textile mill waste, according to Equation (2-4b), is shown in Fig. (2-2), There is a limit to the amount of BOD which can be initially removed by a unit mass of sludge. This limiting removal will be a function of the storage capacity of the

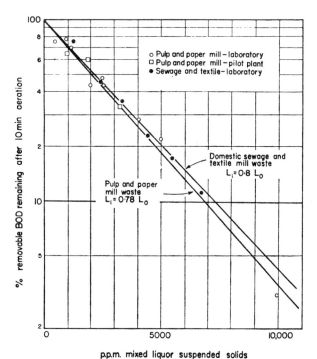

FIG. 2-2. Initial BOD removal characteristics.

cell and of the rate of biological oxidation. A maximum removal by biosorption of 0·65 mg COD/mg VSS from skim milk was found by Porges *et al.* (1955). This removal was stored in the cell as glycogen and was metabolized over a 3 hr aeration period.

To illustrate, consider the bacterial cell in the endogenous phase to have the general composition $C_5H_7NO_2$. Immediately after aeration-contact with organic waste the sludge enters the active respiration-storage phase and its general composition can be gener-

PRINCIPLES OF BIOLOGICAL OXIDATION 19

alized chemically as $C_5H_7NO_2 \cdot C_xH_yO_z$. As aeration continues and the soluble BOD in the waste is removed by the sludge mass, the micro-organisms consume the stored material for metabolism and growth. After sufficient aeration for complete oxidation and synthesis of the BOD removed from the waste, the sludge is again reduced to the endogenous form $C_5H_7NO_2$. (The general formula for sludge ($C_5H_7NO_2$) is that defined by Hoover et al. (1952).)

It is probable that initial removal primarily governs the reduction of BOD through high-rate trickling filters. Velz (1948) and Stack (1957) described the performance of trickling filters in terms of initial removal of BOD by the filter slimes. Their relationships showed that a constant fraction of removable BOD was removed for each unit depth of filter up to a limiting removal. The mathematical relationship developed for the trickling filtration process is similar to that found for activated sludge (see Chapter 6).

The magnitude of the initial removal is dependent on the condition of the sludge when it is contacted with waste. If the sludge has had an insufficient period of aeration prior to contact with waste previously stored BOD will not be completely metabolized and the initial removal will be reduced. Results reported by the Water Pollution Research Laboratory (1956) showed that ½ hr aeration after contact of sewage with sludge resulted in a progressive loss of clarifying power while 2 hr aeration maintained a continued high degree of clarification. On the other hand, excessively long sludge aeration periods result in extensive endogenous respiration and resulting loss in BOD removal capacity of the sludge. Wuhrman (1956) demonstrated a rapid reduction in the removal rate of lactose by activated sludge with sludge aeration periods up to 216 hr. Application of initial removal to activated sludge process design is detailed in Chapter 6.

Mathematical Relationships

From an engineering viewpoint we may consider the various phases of sludge growth and BOD removal to consist of a dynamic relationship between the mass transfer of essential foods into the cell structure, and the assimilation and utilization of these foods for energy and growth. At high concentrations of organic matter the rate of assimilation and the growth rate is independent of the external concentration of organic matter. At low food levels in mixed systems the rate of growth and hence the BOD removal rate are frequently observed to be concentration dependent.

At the beginning of the aeration period let:

S_0 = the initial sludge mass per unit volume (p.p.m.)
L_0 = the total amount of initial BOD that can be oxidized as a limit of the oxidation process.

At any time (t) let:

S = the sludge concentration present (p.p.m.)
$\Delta S = S - S_0$ = the increase in sludge concentration
L_r = the BOD removed
$L = L_0 - L_r$ = the oxidizable BOD remaining
K_1 = logarithmic growth rate for the log growth phase (natural logarithms).
K_2 = logarithmic BOD removal rate when the growth rate becomes BOD concentration dependent (natural logarithms).

Let a be the fraction of the BOD removed which is synthesized to sludge at any time. Then, $aL_r = \Delta S$, and, at any time,

$$S = S_0 + \Delta S = S_0 + aL_r \qquad (2\text{-}5)$$

Phase I—Logarithmic Growth Phase (a-b) on Fig. 2-1

The log growth phase as discussed above may be expressed mathematically:

$$\frac{dS}{dt} = K_1^* S \qquad (2\text{-}6)$$

which in its integrated form is

$$\ln \frac{S_0 + \Delta S}{S_0} = K_1 t \qquad (2\text{-}7a)$$

* K_1 may be interpreted in terms of the generation time of the culture which is the time required to just double the population.

In terms of BOD removal (L_r), Equation (2-7a) becomes:

$$\ln \frac{S_0 + aL_r}{S_0} = K_1 t = \ln\left(1 + \frac{aL_r}{S_0}\right) \quad (2\text{-}7b)$$

The plot of $\ln(1 + aL_r/S_0)$ against t is a straight line function within Phase I (a-b). The slope of this line defines K_1, the constant logarith-

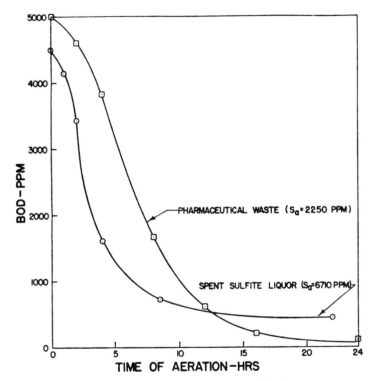

Fig. 2-3. BOD removal characteristics from two industrial wastes.

mic growth rate. The decreasing slope of the line after point b indicates the declining growth rate of Phase II. (This equation directly applies only when there is no storage.) BOD removal through the various growth phases is shown in Fig. 2-3.

The generation times of several bacterial species for conditions of unrestricted growth are shown in Table 2-1. (Buswell *et al*, 1950.)

TABLE 2-1. GENERATION TIMES OF A FEW HETEROTROPHIC ORGANISMS

Organisms	Generation time	
	Minutes	K_1 (per hour)
Proteus vulgaris	21·5	1·94
Escherichia coli	16·5–17	2·48
Aerobacter aerogenes	17·2–17·4	2·40
Eberthela typhosa	23·5	1·77
Diplococcus pneumoniae, type 1	20·5	2·03
Clostridium butyricum	51·0	0·81
Rhizobium trifolii	101–174	0·24–0·41
Rhizobium japonicum	343·8–460·8	0·09–0·12

From studies with glucose and peptone with mixed cultures Garrett and Sawyer (1952) established maximum reaction rates of 0·08/hr at 10°C, 0·20/hr at 20°C and 0·30/hr at 30°C. At 20°C this is equivalent to a generation time of $3\frac{1}{3}$ hr. Maximum growth rates varying from 0·05/hr to 0·13/hr have been found for various industrial wastes treated with mixed cultures.

Within any portion of the log growth curve where the percent increase in sludge mass is not greater than 100 per cent ($\Delta S < S_0$) relatively little error is introduced by letting $S = S_a$, the average sludge concentration over the range under consideration. Then, if $t_0 = 0$, Equation (2-7b) becomes

$$K_1 = \frac{aL_r}{S_a t} \tag{2-7c}$$

Equation (2-7c) can also be used as the more simplified expression for the entire approximately linear portion (e-f) of the growth curve. Equation (2-7c) indicates that the removal rate, expressed in mg BOD/hr/g sludge will be approximately constant. The linear growth phase is shown in Fig. 2-4. The average removal rates for pharmaceutical, brewery, refinery and spent sulfite liquor were found to be 200, 100, 131 and 107 mg BOD removed/hr per g sludge respectively.

Wuhrman (1958) has shown that the rate of removal of simple, specific compounds is usually linear with time and sludge solids

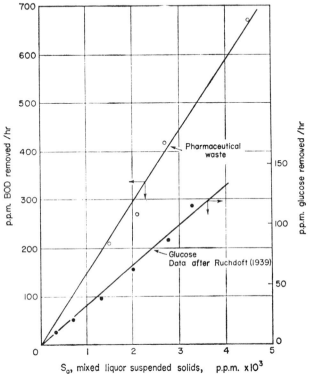

FIG. 2-4. BOD removal relationships during the linear growth phase.

concentration to very low substrate levels (less than 1 p.p.m.). The kinetics of these reactions are explained by the laws governing enzymatic reactions. The removal of phenol by activated sludge appears to follow this linear relationship. The use of non-specific measures of concentration such as BOD, COD, etc., and the complex mixtures of wastes with widely varying reaction rates usually result, however, in a first-order or retardant reaction form.

Phase II—the Declining Growth Phase (b-c) on Fig. (2-1)

At low concentrations of BOD: the sludge growth rate, and hence the BOD removal rate, will frequently be expressed by a first-order reaction. BOD removed under these conditions can be expressed by the relationship. (Eckenfelder and McCabe, 1960):

$$\frac{dL}{dt} = K_2 L \qquad (2\text{-}8)$$

Considering the effect of sludge solids, Equation (2-8) becomes:

$$\frac{dL}{dt} = K_2' S_a L \qquad (2\text{-}8a)$$

and integrating,

$$\ln \frac{L_e}{L} = K_2' S_a t \qquad (2\text{-}8b)$$

where L_e is the BOD remaining at time t.

The heterogeneous nature of most waste mixtures results in a complex removal reaction. The various waste constituents are subject to varying reaction rates, resulting in a decreasing overall rate as the various components are oxidized. In some cases, a small fraction of the BOD does not appear to be removed even after long aeration periods. Weston *et al.* (1960) have attributed this phenomenon to an equilibrium between BOD removal from solution and the release back to solution of the end-products of respiration and auto-oxidation of cellular substance. In cases where this is observed it is frequently necessary to subtract this non-removable BOD from the applied loading to formulate the rate equations.

It is possible to approximate the progress of oxidation of sewage and waste mixtures by employing a composite exponential, which will consider all the various removal rates:

$$L = L_a e^{-(K_2)_a S_a t} + L_b e^{-(K_2)_b S_a t} \ldots + L_n e^{-(K_2)_n S_a t} \qquad (2\text{-}9)$$

in which

$$L_a + L_b \ldots + L_n = L$$

An example of the application of this equation is shown in Fig. 2-5.

PRINCIPLES OF BIOLOGICAL OXIDATION

In Equation (2-9) a high initial removal rate may be reflected in the first term and the decreasing removal rate of the remaining waste constituents in the subsequent terms of the equation. The form of Equation (2-9) was used by Theriault and McNamee (1930) to evaluate the oxidation of channel sludges, by Moore (1937) to evaluate the change in rate of oxygen utilization in activated sludge aeration,

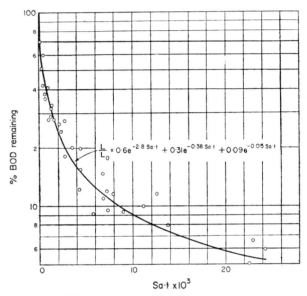

FIG. 2-5. BOD removal characteristics from domestic sewage.

by Streeter (1934) in the evaluation of coliform death rates and by Gameson and Wheatland (1958) in the evaluation of the BOD reaction.

Several other mathematical relationships have been developed to consider the overall reaction rate K_2 to decrease with concentration or time. Fair and Geyer (1954) have employed a generalized expression in which the reaction rate K_2 decreases with a decreasing availability of the substrate:

$$-\frac{dL}{dt} = K_2 \left(\frac{L}{L_0}\right)^n L \qquad (2\text{-}10)$$

which integrates to
$$\frac{L}{L_0} = \left[\frac{1}{1 + nK_2 t}\right]^{1/n}$$

In Equation (2-10) a decreasing reaction rate is reflected by the exponent n. When $n = 0$, Equation 2-10 reduces to Equation 2-8. The removal of BOD from a pulp and paper mill waste (initial BOD

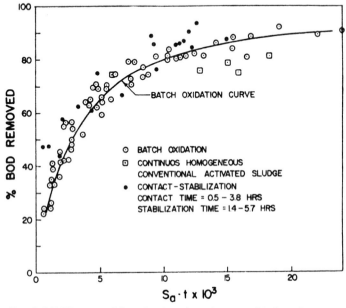

FIG. 2-6. BOD removal from batch and continuous oxidation of a pulp and paper mill waste.

180–250 p.p.m.) during a batch oxidation, which follows this relationship and from a homogeneously mixed pilot plant is shown in Fig. 2-6.

Fair and Moore (1932) have employed a form of equation in which the reaction rate K_2 decreases with time:

$$-\frac{dL}{dt} = \frac{K_2}{(1 + mt)} L \qquad (2\text{-}11)$$

which integrates to
$$L/L_0 = (1 + mt)^{-K_2/m}$$

PRINCIPLES OF BIOLOGICAL OXIDATION 27

In cases where the initial removal rate is low, biological oxidation data can frequently be expressed by Equations 2-10 and 2-11. In cases of variable aeration solids $S_a . t$ may be substituted for t. High initial removal rates will cause a deviation from the equation at low values of $S_a . t$.

Continuous processes. The bio-oxidation of industrial wastes in continuous systems results in short-circuiting of the tank contents due to longitudinal mixing. For any system, three basic mixing concepts have been defined (Greenhalgh *et al.*, 1959), viz. (1) Batch treatment in which the waste flow leaves the aerator in the same order in which it entered. Laboratory oxidation studies are defined by this model. (2) Complete mixing in which the feed completely intermixes with the aeration tank contents and the composition of the effluent is the same as that in the aeration tank. Square or circular aeration tanks with a high degree of agitation would approach this condition. (3) Intermixing in which the feed is uniformly dispersed and each element of feed is in the aerator for a different length of time. Long rectangular aeration tanks would follow this model.

If the oxidation process is in the linear growth phase, or the reaction is zero order, short-circuiting should have little effect on the BOD removal efficiency, since the reaction rate is constant and growth is independent of food concentration. The aeration volume will be the same whether computed for batch flow or complete mixing. When the process is operating in Phase II, however, the removal rate is dependent on the concentration of BOD in the aeration tank at any time. For complete mixing the rate of BOD removal will be equal to $-dL/dt$ at the concentration of BOD in the aeration tank at any time.

In Phase II if the removal rate follows first-order kinetics, Equation 2-8a, it can be shown by a material balance that the BOD removal efficiency in the case of complete mixing will be:

$$\% \text{ Eff.} = 100 \frac{K_2' S_a t}{1 + K_2' S_a t} \qquad (2\text{-}12)$$

When first-order kinetics apply, the process efficiency will vary between that computed from Equations (2-8b) and (2-12), depending on the degree of longitudinal mixing. Fig. 2-6 shows the BOD removal relationship for a batch oxidation and for a completely mixed pilot plant treating pulp and paper mill wastes. It should be noted that in some cases the difference between batch and continuous process may be less than that predicted from Equation (2-8b) and (2-12), due to the higher activity of sludge in a continuous system.

Biological Flocculation

Biological sludge exhibits the tendency to form flocculent masses which settle rapidly under quiescent conditions. Some of the floc-forming organisms found in activated sludge are *Z. ramigera, Esch. intermedium, Para colobactrum aerogenoids, Narcodia actinomorpha, Bacillus cereus* and *Flavo-bacterium* (McKinney and Horwood, 1952). The floc-forming tendency has been indicated by McKinney (1956) to be a function of the growth phase of the culture; the greater the number of living cells the less the floc-forming tendency. In other words, flocculation occurs when the system approaches the endogenous phase (low-loading levels). At high-loading levels sludge does not tend to flocculate and functions as a dispersed growth. Busch and Kalinske (1956) have attributed non-flocculent properties to a young sludge population in the log growth phase.

Various concepts have been offered to explain the mechanism of flocculation. McKinney (1952) has attributed it to electrokinetic phenomena on the bacterial surface and to the chemical composition of the slime layer.

While there is some doubt as to the interpretation of the effects of high loadings on the overall process performance, it is generally conceded that sedimentation and compaction of sludge are impaired when high-loading levels are employed.

SUMMARY OF BIO-OXIDATION KINETICS

The biological oxidation of organic wastes can be assumed to be a 3-phase process as follows:

1. An initial high rate removal of BOD on contact with biologically active sludge. This BOD is stored in the cell as a reserve food source. The extent of this removal depends upon the sludge loading ratio, the type of waste, and the ecological condition of the sludge.

2. Removal of BOD in direct proportion to biological cell growth.

For any organic substrate a portion of this BOD removed is synthesized to new cell material, and the remainder is oxidized to provide energy for this synthesis.

3. Oxidation of biological cell material through endogenous respiration.

The initial removal of BOD occurs in 1–20 min. These data can be obtained by contacting waste and sludge at various loading levels for a predetermined period (usually 15 min) and computing absorption less oxidation and synthesis or by extrapolation of growth data plotted in terms of BOD. The initial removal of BOD is of considerable importance in biological waste treatment since it establishes minimum structural requirements. In domestic sewage treatment, as much as 90 per cent BOD removal has been attained through the initial removal reaction. In the second phase of the process, BOD removal occurs concurrently with synthesis and is related to oxidation, following the log growth-phase law of cell growth. Stored BOD removed on initial contact is subsequently oxidized and synthesized. A material balance relating BOD removal, sludge growth and oxidation may be employed to determine oxygen requirements and excess sludge production for specific process operating conditions.

While endogenous respiration is presumed to occur under all ecological conditions, sludge is destroyed by oxidation when the organic loading is insufficient to support active growth. Extensive endogenous respiration will produce a sludge of low activity and reactive capacity.

Summary of Biological Oxidation Kinetics

Condition	BOD Removal Equation	Remarks
1. High BOD wastes	$2 \cdot 3 \log \left(1 + \dfrac{aL_\mathrm{r}}{S_0}\right) = K_1 t$	Applicable when the initial BOD/sludge solids ratio exceeds 2: applies only when BOD removal is independent of BOD concentration (usually at BOD levels above 200–300 p.p.m.
2. High BOD wastes (Zero order reaction)	$\dfrac{L_\mathrm{r}}{t} = \dfrac{K_1 S_\mathrm{a}}{a}$	Applicable when initial BOD sludge ratio is less than 2. Same limitations as Equation (1).

3. Low BOD wastes $\dfrac{L_r}{L_0} = (1 - e^{-K'_2 S_a t})$ Described course of BOD
   (Zero order reaction)   removal for single substrates with initial BOD's less than 300 p.p.m.

4. $L = L_a e^{-(K_2)_a t} + L_b e^{-(K_2)_b t} + L_b e^{-(K_2)_n t}$

5. $-\dfrac{dL}{dt} = \dfrac{K_2}{(1 + mt)} L$ Equations 4, 5, and 6 estimate BOD removal from complex waste substrates in which the removal rate decreases with time or concentration.

6. $-\dfrac{dL}{dt} = K_2 \left(\dfrac{L}{L_a}\right)^n L$

Example 2-1. Given the following data from the batch oxidation of a pharmaceutical waste:

Aeration time, hr	BOD, p.p.m.
0	4594
½	4072
2	3753
4	3303
8	2240
16½	245
24	90

The initial solids, $S_0 = 1953$ p.p.m. and the coefficient $a = 0.74$

(a). Compute the growth rate K_1
(b). Compute the initial removal
(a) Calculation of the growth ratio:

t	L	L_r	S_0	$\dfrac{aL_r}{S_0}$	$\log\left(1 + \dfrac{aL_r}{S_0}\right)$
0	4594	—	1953	—	—
½	4072	522		0·198	0·078
2	3753	841		0·319	0·119
4	3303	1291		0·488	0·169
8	2240	2354		0·89	0·275
16½	245	4349		1·64	0·421
24	90	4504		1·72	0·434

These data are plotted on Fig. 2-7, from the slope of the plot

$$K_1 = 0.061$$

(This calculation is based on all the stored BOD being consumed at the end of or shortly beyond the log growth phase (Phase I).

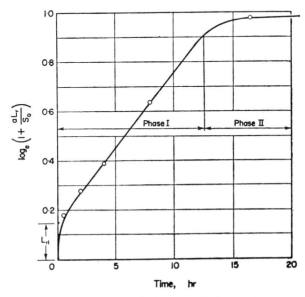

FIG. 2-7. Correlation of pharmaceutical waste data.

(b) Calculation of initial BOD removal (L_{ri})

from Fig. 2-7 when $t = 0$

$$0.06 = \log_e\left(1 + \frac{aL}{S_0}\right)$$

$$0.06 = \log_e\left(1 + \frac{0.74\,L_{ri}}{S_0}\right)$$

$$L_{ri} = 423 \text{ p.p.m.}$$

Example 2-2. An activated sludge process has a K_1 rate of 0·07/hr. If the BOD is 5000 p.p.m., what aeration volume will be required to attain 90 per cent BOD reduction for a flow of 1 Mgal/day? The initial MLSS is 3000 p.p.m. and the return sludge concentration is 20,000 p.p.m. (assume a = 0·70). Neglect influent solids.

From a solids material balance

$$r = \frac{R}{Q} = \frac{S_a}{S_r - S_a}$$

$$r = \frac{S_a}{S_r - S_a} = \frac{3000}{20{,}000 - 3000} = 0\cdot176$$

$$l_0 = \frac{Q\,l_a + R\,l_e}{Q + R} = \frac{(1)(5000) + (500)(0\cdot176)}{1\cdot176} = 4320 \text{ p.p.m.}$$

$$l_r = 4320 - 500 = 3820 \text{ p.p.m.}$$

then from Equation (2-7b)

$$t = \frac{2\cdot3 \log\left(1 + \frac{(0\cdot7)(3820)}{3000}\right)}{0\cdot07}$$

$$= 9\cdot1 \text{ hr}$$

$V = 1/24 \cdot 9\cdot1 \text{ hr} = 0\cdot38 \text{ mg}$

Example 2-3. It is desired to treat 3 Mgal/day of a waste with an initial BOD (l_a) of 250 p.p.m. The required BOD reduction is 80 per cent. If the BOD removal follows the relationship as shown in Fig. 2-6, compute the volume of the aeration tanks required.

Design Data
    Settled sludge return $S_r = 12{,}000$ p.p.m.
    Mixed liquor suspended solids $S_a = 3000$ p.p.m.

Sludge recycle ratio:

From a material balance of the process (neglecting solids in the influent waste) the sludge recycle ratio ($r = R/Q$) becomes:

$$r = \frac{S_a}{S_r - S_a} = \frac{3000}{12{,}000 - 3000} = 0\cdot33$$

BOD at beginning of the aeration period:

$$l_0 = \frac{Q\,la + R\,le}{Q + R}$$

$$= \frac{(3)(250) + (1)(50)}{4} = 200 \text{ p.p.m.}$$

If effects of short-circuiting are neglected:

per cent BOD removal (based on l_0) = $\frac{150}{200}$ = 75 per cent

and from Fig. 2-6

$$S_a \cdot t = 7800$$

For $S_a = 3000$ p.p.m., $t = 2.6$ hr
Volume of aeration tank

$$V = 4 \text{ Mgal/day} \times \frac{2.6}{24} = 0.433 \text{ mg}$$

BIO-OXIDATION OF PURE COMPOUNDS

Studies on the oxidation of various pure organic substances have been conducted by several investigators. Some of their conclusions are summarized herein.

Normal alcohols above methanol undergo oxidation, through aldehydes, and organic acids to carbon dioxide and water. Some of the iso-alcohols undergo similar oxidation as the normal alcohols and some are retarded. The 1 and 3 carbon alcohols are less readily oxidized than the 2 and 4 carbon alcohols. Sludges acclimatized to alcohols can oxidize related compounds (McKinney and Jeris, 1955).

Formaldehyde (Gellman and Heukelekian, 1950) is oxidized by a pink flocculent growth of *Bacterium methylicum*. This sludge settles rapidly on standing. Ninety-five per cent BOD reduction is attained with concentrations up to 1750 p.p.m. formaldehyde in 24 hr or less aeration. The maximum observed oxidation rate was 190 p.p.m./hr.

The minimum BOD : N ratio is 40 : 1. A minimum operating pH of 6·0 was found.

Ninety-nine per cent BOD reduction can be attained from the oxidation of oxalic acid at feed levels up to 2000 p.p.m. (83 lb/1000 ft^3/day at a pH of 2·3–3·5 (Nelson *et al.*, 1954). Sludge acclimatized to phenol can oxidize *O*-cresol, *m*-cresol, *p*-cresol, benzoic acid and benzaldehyde in concentrations up to 500 p.p.m. (McKinney *et al.*, 1956).

Within variable limits, all the principal anionic and nonionic detergents are susceptible to some degree of bio-oxidation (Bogan and Sawyer, 1955). The alkylsulfate, sulfonated fatty acid amides and esters and derivatives of low molecular weight polyethylene glycols undergo rapid oxidation while the alkylaryl sulfonates, alkylphenoxy polyglycols and derivatives of high molecular weight polyethylene glycols undergo slow oxidation. The theory offered is that only the smaller ether molecules can reach the enzymes centers.

CHARACTER OF BIOLOGICAL SLUDGE

The microbial sludge or biological floc employed in bio-oxidation processes is a miscellaneous collection of micro-organisms such as bacteria, yeasts, molds, protozoa, rotifers, worms and insect larvae in a gelatinous mass. Algae will also be present in those areas exposed to sunlight.

An excellent compendium of the ecology of activated sludge and trickling filters has been presented by Hawkes (1960). The bacteria are primarily non-nitrifying aerobic spore formers, many of which are of the *B. subtilis* group. Nitrifying bacteria are primarily *Nitrosomonas* and *Nitrobacter*. In most activated sludge processes the sludge appears as zoogleal masses intermixed with filamentous bacteria. One of the principal forms in the zoogleal mass is *Zooglea ramigera* which has been defined as a gram-negative, non-spore forming, motile, capsulated rod. Since most bacteria under proper conditions can flocculate, *Zooglea ramigera* may not be a true species, but rather a growth form of many species. Other common forms of bacteria found in activated sludge include *Flavobacterium*, *Pseudomonas* and filamentous organisms, the most common of which is *Sphaerotilus natans*. Fungi are more common in trickling filters than in activated sludge. These forms generally exist in the presence of low oxygen tension, low pH or low nitrogen content. Of the protozoa, stalked ciliates are the most common, including *Vorticella*, *Opercularia* and *Epistylis*. Free swimming types include *Paramoecium*, *Linnotus* and *Trichoda*. Some forms of *Flagellata* and *Rhizopoda* are also found. The relationship between the type of protozoa which predominates and the bacterial population seems to depend on the degree of flocculation. In a well flocculated sludge, stalked ciliates and attached forms are common since they feed on the zoogleal mass. With low flocculation, free swimming forms dominate. The interrelationship between bacteria and protozoa on treatment efficiency is not well defined. It is generally conceded, however, that protozoa aid in clarification. Englebrect and McKinney (1957) found that sludge developed on structurally related compounds possessed similar morphological characteristics and produced similar biochemical changes. For example, they found that dense flocs were produced from the pentose sugars xylose and arabinose and that filamentous floc was produced from the hexose sugars glucose and fructose.

A wide variation in micro-organism population between the assimilative and the endogenous phases was reported by Jasewicz and Porges (1956) on dairy waste studies. They found that during the assimilative phase, 74 per cent of the organisms were of the genera *Bacillus* or *Bacterium* while only 8 per cent of the endogenous sludge was composed of these organisms. The endogenous sludge contained 42 per cent of the proteolytic organisms *Pseudomonas* and *Alcaligenes* and 48 per cent of the sacchorolytic organisms *Flavobacterium* and *Micrococcus*. Protozoan forms were similar in both sludges. The distribution of micro-organisms in trickling filters, likewise, varies with depth depending on the food supply and the growth conditions. In general, algae forms are found in the surface layers and a predominance of nitrifying forms in the lower depths. Activated sludge from two bio-oxidation processes is shown in Fig. 2-8.

The chemical content of micro-organisms will depend upon the quantity of water absorbed by the cell and will be a function of the pH and the environment. Capsulated organisms retain more moisture than those which do not have a capsule. The average moisture content of bacteria is 80 per cent (73–88 per cent). The moisture content of yeasts averages 75 per cent and molds 85 per cent. Phosphorus, potassium, magnesium and other trace minerals are present in the ash or mineral content and are partly organically bound in proteins, nucleic acids, carbohydrates, lipids, pigments, etc. A high proportion of the total ash consists of potassium and phosphorus. The variation in phosphorus content for most micro-organisms is 2·5–5·0 per cent. The ash content of most biological sludges is low (2–15 per cent). When the ash content exceeds this, it is generally due to the presence of non-biological inert substances in the sludge. Nitrogen is present in protoplasm as proteins and amino acids (product of the breakdown of proteins and nucleo-proteins). Nitrogen content will vary from 8–15 per cent for most bacteria. Yeasts and molds will possess a lower nitrogen content. The nitrogen content is measured by the Kjeldahl distillation. Protein content may be computed as the total organic nitrogen measured above times 6·25. This factor is based on an average nitrogen value of 16 per cent for protein. The carbon content of cells is present as complex carbohydrates, etc. and is measured by converting the dry mass to CO_2 and H_2O in which the carbon is equal to 3/11 of the CO_2. The hydrogen content is present as 1/9 of the H_2O produced by the conversion. The carbon content of most micro-organisms ranges from 45–55 per cent. Lipoids (fat) are also present to a varying degree in the cell mass.

It is convenient to express the composition of a cell mass in terms of mean constituents and to derive an empirical formula which expresses the statistical average properties of the major atoms of the organic molecules. The following table shows these data for typical organisms found in bio-oxidation processes.

Hoover and Porges (1952) showed sludge synthesized from dairy wastes to have the general empirical formula $C_5H_7NO_2$. This formula is representative of the ratio of the primary element constituents of activated sludge. It is representative of the statistical average composition of the complex organic compounds constituting cell material. This sludge contains 12·3 per cent nitrogen and theoretically requires 1·42 g oxygen for each gram of sludge completely oxidized. Hoover found an empirical factor of 1.25 g O_2 per g of dry weight solids for dairy waste oxidation. This is equivalent to 1·36 g O_2 per g dry weight volatile solids for a sludge of 8 per cent ash.

TABLE 2-2. ANALYSES AND EMPIRICAL FORMULAS OF MICRO-ORGANISMS

	Yeast*	Bacteria†	Zooglea
Carbon	47·0	47·3	44·9
Hydrogen	6·0	5·7	—
Oxygen	32·5	27·0	—
Nitrogen	8·5	11·3	9·9
Ash	6·0	8·3	—
Empirical formula	$C_{13}H_{20}N_2O_7$	$C_5H_7NO_2$	—
Carbon–Nitrogen ratio	5·6/1	4·3/1	4·5/1

* Porges (1953).
† Hoover and Porges (1952)

Studies by Helmers et al. (1951) on various industrial wastes produced a sludge of 1·02 per cent phosphorus content and 8 per cent nitrogen content with the general formula $C_{118}H_{170}O_{51}N_{17}P$. Ignoring the phosphorus content, this sludge has the formula $C_7H_{11}NO_3$, and requires 1·53 g of oxygen for each gram of sludge oxidized.

Variations in the empirical composition will depend on quantity and availability of nitrogen and phosphorus in addition to other microbiological variables. Studies on several sludges synthesized from various industrial wastes revealed a range of 1·36 to 1·44 g of oxygen per gram of sludge volatile solids oxidized.

The volatile content of activated sludges will depend upon the percentage of inert and non-oxidizable volatile solids accumulated in the system in addition to the growth phase of the microbial culture. Pure microbial cultures undergoing active growth average 90–94 per cent volatile. As oxidation under starvation conditions progresses, this volatile content will decrease due to the mineralization of cellular constituents. Using synthesized milk waste Hoover and Porges (1952) produced a sludge of 92 per cent volatile content. Sludges synthesized from cotton kiering, rag rope and brewery waste showed a variability of volatile content of 72–89 per cent (Helmers et al., 1951). Sludge from pulp and paper waste oxidation was 85 per cent volatile. Most domestic sewage activated sludges are 75–90 per cent volatile.

The volatile solids content is not necessarily indicative of the active fraction. Active fraction in this instance is defined as the fraction of microbial sludge as related to the total sludge weight. For example, in the pulp and paper sludge cited above, while the volatile content was 85 per cent, the active sludge fraction as determined by system material balances, was only 70 per cent. This difference can be attributed to the accumulation of fibre, lignin and cellulose of low oxidation rate in the system.

Oxygen uptake rate will yield information on the active portion of the sludge when the oxygen uptake characteristics of the pure sludge are known. McKinney (1960) has employed the organic nitrogen content of the biological solids as an indication of the active content. A method was developed by Stern (1959) in which a given weight of activated sludge was innoculated into nutrient broth and the increase in cell weight measured by centrifuging and weighing after various periods of aeration. Extrapolating the log growth curve back to zero time gave the true initial biological sludge mass which, in turn, when divided by the initial gross weight, gave the active sludge fraction.

The heat content of activated sludges from industrial waste oxidation have been shown to be 10,321 B.t.u./lb for dairy waste sludge and 9872 B.t.u./lb for pulp and paper waste sludge. This compares to the value for domestic sewage activated sludge of 10,000 B.t.u./lb found by Fair and Moore (1935). The sludge age appeared to have only a minor influence on the heat content. The heat content and other characteristics for a dairy waste sludge are summarized in Table 2-3 (from Weston and Eckenfelder, 1955).

TABLE 2-3. CHARACTERISTICS OF ACTIVATED SLUDGE
(DRY SOLIDS BASIS)

N %	Vol. (%)	Ash %	Heat value (BTU)		Harvest time after feeding (hr)
			Per lb	Per lb Vol. Sol.	
9·49	91·44	8·56	9,369	10,246	24
10·34	90·79	9·21	9,370	10,321	6

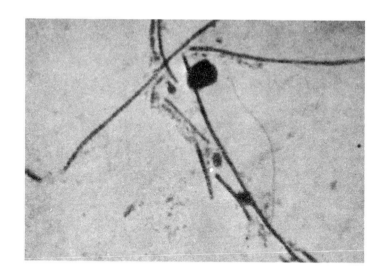
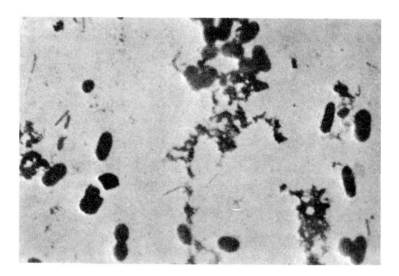

Fig. 2-8. Activated sludge.

 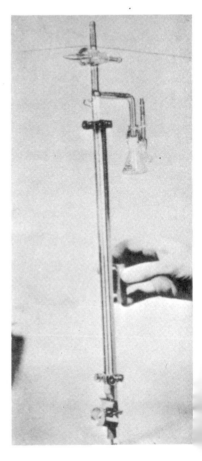

FIG. 2-11 (i). FIG. 2-11 (ii).

FIG. 2-11. Warburg apparatus (*Courtesy of Bronwill Corp.*).

PRINCIPLES OF BIOLOGICAL OXIDATION 39

OXYGEN UTILIZATION

As has been shown in Equations (2-1) and (2-2) oxygen plays an essential role in aerobic biological treatment. For optimum efficiency, oxygen must be supplied at a rate equal to or greater than its rate of utilization. In the activated sludge process this is usually accomplished by diffusion from air bubbles injected into the liquid-sludge mass under turbulent conditions. In trickling filters, oxygen is derived from air drawn into the bed due to the temperature gradient between the waste and the ambient air and from oxygen dissolved in the incoming waste.

Oxygen utilization rate may be defined as the weight of oxygen consumed by a given weight of microbial sludge per unit of time. It is usually expressed as p.p.m. per hour. In some cases the utilization rate may be expressed in other forms. Some of the more common are listed below.

$$r_r = \text{p.p.m. } O_2/\text{hr}$$

$$k_r = \text{mg } O_2/\text{hr/g sludge}$$

$$Q_{O_2} = \text{ml}/O_2 \text{ hr/g sludge}$$

(1 ml of O_2 contains 1·43 mg O_2)

A linear relationship will exist between sludge concentration and oxygen utilization over the range of sludge concentrations usually employed.

$$r_r = k_r S_a \qquad (2\text{-}13)$$

In very high sludge concentrations ($>$10,000 p.p.m.) the unit rate of oxygen utilization may decrease due to diffusional resistances (Dawson and Jenkins, 1949). In most cases the specific uptake rate (k_r) will vary inversely with organism size. The most satisfactory index for specific oxygen uptake rate is the surface area per unit volume (Gaden, 1956).

During assimilation, micro-organisms require oxygen to supply energy required for synthesis. In addition to the above oxidation, the sludge produced by the assimilation of organic matter is continually oxidized by its own mass. Hoover and Porges (1952) have defined this as endogenous respiration. In the absence of available nutrients, cells oxidize their own tissue slowly in order to obtain energy for maintenance. Recent investigations have shown that endogenous respiration also occurs concurrently with synthesis. This oxidation is defined by Equation (2-3).

The total oxygen requirements of a bio-oxidation system can be defined (Eckenfelder and O'Connor, 1954):

$$\text{lb } O_2/\text{day} = a' \text{ lb BOD}_5 \text{ removed/day} + b' \text{ lb MLVSS} \quad (2\text{-}14)$$

The coefficient, a', represents that fraction of the 5-day BOD removed which is used to provide energy for growth. The coefficient,

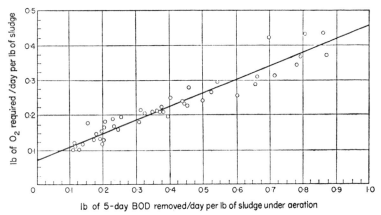

FIG. 2-9. Relationship between oxygen utilization and BOD removal.

b', represents the endogenous respiration rate. Data from a plant treating pulp and paper mill wastes are shown in Fig. 2-9. In a variety of industrial waste oxidation systems the coefficient a' has been found to vary from 0·35–0·55. In order to complete a material balance with sludge growth according to Equation (2-2) a' must be converted to terms of ultimate BOD.

It is significant to note that the availability of nutrients will influence the coefficient a'. Wuhrman (1956) found that only 16 per cent of glucose sugar was respired by washed activated sludge in the absence of nitrogen while 50 per cent was respired

PRINCIPLES OF BIOLOGICAL OXIDATION 41

when nitrogen was present. These observations were confirmed by Symons and McKinney (1958) in which oxidation in the presence of low nitrogen resulted in a large accumulation of sludge of high polysaccharide content, and hence a lower fraction oxidized.

The maximum oxygen uptake rate encountered in a waste oxidation system will be related to the sludge growth rate and to the endogenous respiration rate. From the growth relationships shown in Fig. 2-1, the maximum uptake rate will be

$$\frac{dO_2}{dt} = \left(\frac{a'}{a} K_1 + b'\right) S \qquad (2\text{-}15)$$

In the declining growth phase, the growth rate K_1 decreases progressively with decreasing substrate levels. The uptake rate computed from Equation (2-15) will decrease and approach the endogenous level as K_1 approaches zero. Observed oxygen uptake rates for various bacterial cultures and waste oxidation systems are tabulated in Table 2-4.

Example 2-4. Compute the total oxygen requirements and the oxygen uptake rate for a bio-oxidation system in which operating data satisfies the relationship:

lb O_2/day = 0·384 (lb $BOD_{removed}$/day)
$\qquad$ + 0·07 lb aeration volatile sludge

lb $BOD_{removed}$/day $\quad = 11{,}250$

aeration volatile sludge $= 2100$ p.p.m.

aeration volume $\quad = 0\cdot79$ mg

lb O_2/day = 0·384 (11,250)
$\qquad$ + (0·07) (2100) (8·34) (0·79) = 5300

The oxygen uptake rate is computed:

$$r_r = \frac{\text{lb } O_2/\text{day}}{V\,(8\cdot34)\,(24)} = \frac{5300}{(0\cdot79)\,(8\cdot34)\,(24)} = 34 \text{ p.p.m./hr}$$

Endogenous Respiration

During the early phase of endogenous respiration the rate equation may approximate first-order kinetics and the rate of sludge oxidation may be conveniently expressed as a per cent per day of the sludge solids under aeration. This would correspond to a constant rate of oxidation per unit weight of sludge. Actually the oxidation rate

TABLE 2-4

Waste	Respiration	Rate–mg O_2/hr/g sludge
Sewage	active	10–20
Sewage	endogenous	1·85–9·8
Dairy	active	40–45
Dairy	endogenous	4–10
Cannery	active	35
Pharmaceutical	active	76 (av. max. recorded)
Pulp and Paper	active	10–15

ENDOGENOUS RESPIRATION OF COMMON MICRO-ORGANISMS*

Measurement		k_r mgO_2/hr/g	Organism
Temp. (°C)	Time (hr)		
22	2	17	*Pseudomonas aeruginosa*
22	2	10	*Serratia marcescens*
22	2	23	*Pseudomonas fluorescens*
22	2	10	*Alcaligenes faecalis*
22	2	7	*Proteus vulgaris*
25	2·5–4·5	36	*Bacillus cereus*
25	24	26–11†	—
22	2	6	*Escherichia coli*
25	2·5–4·5	11	—
22	2	7	*Mycobacterium phlei*
22	4	17	*Bacillus subtilis*
22	2	9	*Bacillus Megatherium*
22	2	14	*Micrococcus pyogenes var. aureus*
22	2	14	*Sarcina aurantiaca*
25	2	17†	*Saccharomyces cerevisiae*
37·5	4	7	*Mycobacterium tuberculosis*

* After Hoover and Porges (1953).
† Decreases exponentially with time.

SPECIFIC OXYGEN UPTAKE RATES (k_r) FOR VARIOUS ORGANISMS‡

Organism	Temperature °C	k_r mg O_2/g (hr)
Luminous bacteria	20	11·2
Azotobacter chroococcum	10	300
	30	13·1–250
Escherichia coli	30	12·8
Streptomyces griseus	27	16–48
Paramecium sp.	20	1·1
Euglena gracilis	25	3·2
Chlorella pyrenoida		
endogenous	25	1·4
exogenous	25	11·2
Neurospora sp.		
spores, dormant	?	0·2
spores, germinating	?	12·8
Yeast	30	152
Yeast	25	43
Planaria agilis (worms)	20	0·24

‡ After Gaden (1956).

decreases with time due to the fact that the cell constituents differ in ease of oxidation. It has been shown that this rate of decline for many micro-organisms is logarithmic in nature. As the oxidation of sludge proceeds, cellular nitrogen is broken down and released to solution in the form of ammonia. Under continuous oxidation this ammonia may be further oxidized to nitrates.

A portion of the cellular constituents are highly resistant to oxidation and result in an accumulation of solids in the process. McKinney (1960) and Kountz et al. (1959) have indicated that the non-oxidizable solids buildup from the auto-oxidation of biological sludge may amount to 25 per cent of the sludge produced by synthesis. If non-oxidizable solids are present in the incoming waste the solids buildup will be greater.

The endogenous respiration rate of domestic sewage activated sludge has been reported to vary from 1·9 to 9·8 mg O_2/hr g of sludge undergoing aeration for 24 hr with no nutrient addition (Sawyer and Nichols, 1939). In an activated sludge plant treating semi-chemical and pulping wastes the endogenous respiration rate varied from 3 to 7 mg O_2/hr g sludge (volatile solids basis) over a 10 hr aeration period at 32°C. The BOD loading averaged 0·4 lb BOD/day lb sludge.

In studies on sludge oxidation from kraft mill wastes Gehm (1953) found a mean endogenous respiration rate of 2·1 mg O_2/hr g volatile suspended solids. Aerating a 1 per cent sludge he found a reduction in BOD of 82 per cent after 5 days aeration.

Variations in Oxygen Uptake Rate

In addition to computing the total oxygen requirements for a system, it is important to determine the distribution of the oxygen demand in the aeration tanks in order to design the aeration system. As can be implied from Fig. 2-10, the oxygen uptake rate will vary with time of aeration as the sludge passes through the various growth phases. Unless the entire process occurs in the log growth phase, the growth rate, and consequently the oxygen uptake rate, will decrease along the length of the tank as the BOD to sludge ratio decreases, until the endogenous level is reached.

The oxygen uptake rate will vary with time of aeration as the sludge passes through the various growth phases depending on the concentration of sludge employed and the BOD of the waste. If the BOD to sludge ratio (L_0/S_0) is low, the initial rate will be high since rapid oxidation of the BOD is occurring. Due to the rapid initial removal of BOD under these conditions the uptake rate will decrease very rapidly and approach the endogenous level. An example of this was provided by the bio-oxidation of pulp and paper mill wastes with a BOD to sludge ratio of 0·1. The oxygen uptake rate at the start of the aeration period was 37·5 mg O_2/hr g of sludge. After 3 hr aeration the oxygen uptake rate was reduced to 7·4 mg O_2/hr g of sludge. Conversely, if the BOD to sludge ratio is large, a high rate will be maintained for a long period of time as the BOD removal and oxidation process continues.

Any aeration tank will have some degree of mixing and short-circuiting. In the long, rectangular aeration tanks commonly employed for the activated sludge process, longitudinal mixing will dampen the uptake rate variation through the tank length. In circular or square aeration tanks with a high degree of turbulence, the tank contents approach a homogeneous condition and the uptake rate is a constant in all portions of the tank. In order to design an aeration system it is necessary to know something of tank geometry and turbulence characteristics.

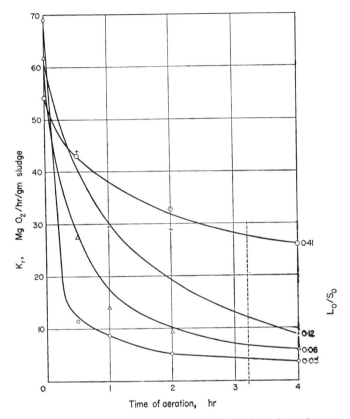

FIG. 2-10. Variation in oxygen utilization with time of aeration.

Oxygen Concentration

It has been found that when the oxygen concentration in the mixed liquor is greater than 0·2–0·5 p.p.m., the rate of bacterial respiration is independent of oxygen concentration. When the oxygen concentration is below this value, the system becomes oxygen dependent and the rate of BOD removal is decreased.

When an aeration process is operated below the minimum oxygen concentration, the sludge will undergo degradation which will subsequently impose an oxygen demand on the aeration system. For example, in a plant treating pulp and paper mill waste, the oxygen uptake rate in the first quarter of the aeration tanks was 26 p.p.m./hr at dissolved oxygen levels of 1·0 p.p.m. or greater. When the sludge was subjected to oxygen values of less than 0·5 p.p.m. for a day or more the oxygen uptake rate rose to 64 p.p.m./hr.

Sludge cells tend to clump, hence to decrease the quantity of oxygen which can be transferred to them by an increase in resistance to transfer. Warburg (1950)* and Pasveer (1954) defined mathematical relationships for oxygen diffusion into microbial cells which are a function of floc size, diffusivity, oxygen utilization rate and external concentration of dissolved oxygen (driving force). A relationship can be developed incorporating these variables for the case when the entire floc contains oxygen. The oxygen transferred to the sludge floc from the surrounding liquid can be approximated by Equation (2-16a)

$$\frac{dM}{dt} = \frac{D}{R} \cdot 4\pi R^2 (C_L - C_M) \qquad (2\text{-}16a)$$

in which

$\dfrac{dM}{dt}$ = weight rate of oxygen transfer

C_L = oxygen concentration at the cell interface

C_M = oxygen concentration within the cell

D = diffusivity of oxygen

R = mean floc radius

(Assuming the floc is spherical, the change in rate with decreasing floc radius is neglected.)

* See Hober *et al.* (1950).

The oxygen consumed by the sludge floc will be:

$$\frac{dM}{dt} = k_r \cdot \rho \cdot \frac{4}{3} \pi R^3 \qquad (2\text{-}16b)$$

where ρ is the floc density (1·018–1·21)
For steady state conditions Equation (2-16a) = Equation (2-16b)

$$k_{r\rho} 4/3 \pi R^3 = \frac{D}{R} 4\pi R^2 (C_L - C_M) \qquad (2\text{-}16c)$$

and

$$R = \sqrt{\frac{3D(C_L - C_M)}{k_r \rho}} \qquad (2\text{-}16d)$$

In the case of trickling filters oxygen transfer occurs across a plain interface at the film surface and the effective film depth may be expressed (see Chapter 6):

$$h = \sqrt{\frac{D \, C_L}{k}} \qquad (2\text{-}17)$$

Wuhrman (1960) estimates the maximum depth of oxygen penetration in conventional filters as 100–200 μ. The diffusion coefficient, D, for oxygen into cell material has been estimated as 5×10^{-6} cm^2/sec at 15°C (Wuhrman, 1960). Wuhrman also indicated that oxygen concentrations in the center of the floc of 0·1 p.p.m. may be adequate. From these considerations he deduced that for conventional sewage treatment plants at concentrations of 1·5–2·5 p.p.m. dissolved oxygen, maximum floc diameters of 400–500 μ can be supplied with oxygen. In Equation (2-17) it is assumed that the concentration at the interface is saturation such that $C_L \equiv C_s$ at the temperature of the flowing waste.

High degrees of agitation will disperse the sludge clumps and increase the transfer rate to the cells for metabolism. By decreasing the mean floc radius, a greater surface is exposed for oxygen transfer and the degree of oxygen penetration is increased. This has the net result of increasing the unit rate of oxygen utilization. The maximum turbulence in the system will be limited by that power input which will not excessively shear the floc particles for subsequent settling. Turbulence and power input to the aeration system is frequently expressed as horsepower absorbed per 1000 gal of tank capacity.

Measurement of Oxygen Utilization Rates

Oxygen utilization rates may be measured (a) by the direct absorption of gaseous or dissolved oxygen or (b) by the indirect method of measuring the drop in the oxygen demand of a sludge. The direct method is applicable to whole suspensions or resuspensions of cells in buffer, or special solutions.

Manometric Analysis

In the direct method a sludge sample is respired in a closed oxygen or air atmosphere at constant temperature. The sample is agitated and the oxygen utilized is measured with respect to time by noting the decrease in gas volume or pressure. The Sierp and Warburg assemblies are typical examples. The Warburg apparatus consists of a reaction flask connected to a manometer in a constant temperature bath. The sludge-waste mixture is kept agitated by a shaker assembly. As oxygen is biologically utilized it is replenished from the gaseous phase above the sample. The system is maintained at constant volume by adjusting the manometer columns before reading. The CO_2 evolved is eliminated by absorption in KOH in a small center well of the test flask. The removal of oxygen from the gas phase creates a pressure difference which may be read on a manometer. A control flask is employed to correct for barometric and temperature variations. As Gaden (1956) has shown, one drawback to this procedure lies in the fact that the measured rate is due to a decrease in oxygen partial pressure in the gas phase rather than of oxygen in solution. With high uptake rates, oxygen absorption from gas to liquid may limit the respiration rather than the characteristics of the culture itself. This method has been employed by Grant, Hurwitz and Mohlman (1930) and others. The Warburg apparatus is shown in Fig. 2-11. The formulae and constants necessary to compute oxygen utilization from Warburg data are summarized below:

$$\text{p.p.m. } O_2 \text{ utilized} = \frac{1000}{V} \frac{32}{22 \cdot 4} hk = 1430 \frac{hk}{V}$$

where V = volume of sample, ml

h = manometer pressure change, cm

k = a flask constant determined for each flask, sample, volume and temperature by the following formula:

$$k = \frac{V_g \times \frac{273}{T} + V_f d}{P_0}$$

V_g = gas volume in closed system, ml

V_f = liquid volume in flask, ml

$d = \dfrac{0.0325 \text{ ml gas}}{\text{ml liquid}}$ = mg O_2 per ml liquid, when total pressure (the sum of the partial pressure of the gas plus the aqueous tension 20°C) is 760 mm Hg and temperature is 20°C

P_0 = 1·001 cm (1 atm of Brodie's solution)

T = absolute temperature, °K

It should be emphasized that the Warburg yields a summation of oxygen utilization. The instantaneous utilization rate is obtained from the slope of the oxygen-time curve.

Warburg data may be advantageously employed to evaluate the oxidation characteristics of industrial wastes and to determine toxicity effects. Some typical curves are shown in Fig. 2-12.

Polarographic Analysis

Recently a polarographic technique for oxygen utilization rate has been developed (Hixon and Gaden, 1950). Dissolved oxygen is measured by currents produced by the reduction of oxygen at a continually renewed mercury or platinum electrode surface. At a specified applied voltage, usually 0·5–0·6 V vs. the standard calomel electrode, the current produced is proportional to the concentration of dissolved oxygen. By measurement of the decrease in oxygen content with respect to time, the rate of oxygen utilization by the micro-organisms may be computed.

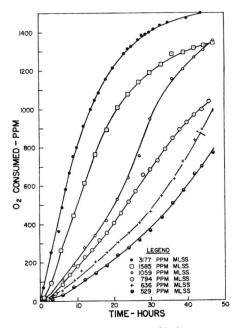

FIG. 2-12. Typical Warburg oxidation curves.

The sludge sample is air saturated by shaking and the oxygen concentration-time curve plotted. The curve is linear to low residual oxygen values (0·1–0·3 p.p.m.). The slope of the curve is the uptake rate, r_r, and may be converted to the unit rate k_r if the sludge concentration is known.

Procedure: Oxygen Utilization Rate:
(1) The sludge sample is completely saturated with oxygen by agitating in a stoppered Erlenmeyer flask for a period of 30 to 60 sec or aerating for 1 to 3 min. It is then placed in the polarograph cell.

(2) As soon as the diffusion current comes to a stabilized value (that is, a linear decrease in diffusion current with time), a reading is taken and a timer is started. Care must be exercised to obtain this initial reading. Since the oxygen is being utilized by the bacteria at a rapid rate, the readings must be obtained as soon as the diffusion current becomes stable.
(3) Readings should be taken every half minute or full minute. If working at half-minute intervals, the sample is stirred gently for 10 sec immediately following the readings to avoid sludge deposition. If working at full-minute intervals, the sample is stirred gently for 20 sec immediately following the readings. In order to avoid oxygen entrainment into the sample a nitrogen blanket over the sample is desirable.
(4) About five readings should be taken and then the temperature of the sludge obtained.
(5) The current readings are multiplied by the respective current multipliers and a plot of current or dissolved oxygen vs. time in minutes is made on arithmetic paper (Fig. 2-13).
(6) Oxygen uptake (p.p.m. per hr)

$$= \frac{\Delta D_T}{\Delta t} [1 + 0.014(20 - T)] \times 60 \times f$$

where:

$\dfrac{\Delta D_T}{\Delta t}$ = slope of curve,

T = temperature of sample, °C,

60 = conversion of time, t, to hr, and

f = p.p.m. dissolved oxygen per unit galvanometer deflection.

Oxygen utilization rate determined by this method only maintains the sludge in suspension during the test. If high turbulence levels occur in the treating unit from which the sludge is withdrawn the utilization rate as measured by this test may be low with respect to what actually occurs in the tank itself. This procedure may also be employed to determine the instantaneous oxygen content of aeration tanks by extrapolating the oxygen concentration-time curve to zero time.

Carbon Dioxide Evolution

When the respiratory quotient (CO_2/O_2 ratio) is approximately 1·0, the rate of CO_2 output of the sludge may be used as a parameter

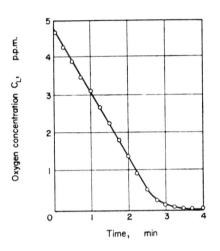

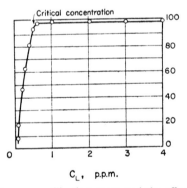

Fig. 2-13. Typical oxygen utilization curve and the effect of oxygen concentration on respiration (after Gaden, 1956).

of oxygen consumption. In the oxidation of most sewages and wastes this value is closely approximated. A simple apparatus has been developed by Porges *et al.* (1952) for CO_2 output studies. The CO_2 evolved from sludge–waste mixture is measured by passing CO_2

free air through the sludge-substrate mixture. The CO_2 laden air from the system is bubbled through a barium hydroxide solution which absorbs the CO_2 as insoluble $BaCO_3$. Excess barium hydroxide is titrated with standardized oxalic acid and the CO_2 computed by difference. A good correlation was obtained by Porges between results obtained with this assembly and Warburg respirometer studies.

Oxygen Gas Analysis

The oxygen consumed by a respiring sludge can be obtained by measuring the change in oxygen content of the aerating gas (gas in minus gas out).

In principle, this is the ideal method, since it measures oxygen absorption under conditions of actual operation and includes effects of turbulence and fluid shear on the sludge floc. The principle drawback is that for the small measurable change in oxygen content through an aeration tank (usually only 5-10 per cent absorption) large errors may result from calculation of the measured values. Details of this method have been described by Hoover *et al.* (1954).

SLUDGE PRODUCTION AND OXIDATION

It has been previously shown from kinetic and material balances that removal of BOD results in the growth of sludge. The amount of sludge growth has been shown to vary with the nature of the sub-

TABLE 2-5. CONVERSION OF ORGANIC COMPOUNDS TO SLUDGE

Substance	Per cent conversion
Carbohydrate	65-85
Alcohols	52-66
Amino acids	32-68
Organic acids	10-60
Skim milk solids	50-52
Glucose*	49-59
Glucose	44-64
Sucrose	58-68

* Yeast.

TABLE 2-6. SLUDGE PRODUCTION FROM PURE COMPOUNDS

Compound	Solids accumulated (vol.) BOD removed	Reference
Methanol	0·16	Gellman and Heukelekian (1955)
Ethanol	0·31	Gellman and Heukelekian (1955)
Iso-propanol	0·09	Gellman and Heukelekian (1955)
n-Butanol	0·28	Gellman and Heukelekian (1955)
Maltose	0·534	Placak and Ruchhoft (1947)
Lactose	1·140	Placak and Ruchhoft (1947)
Lactose	0·52	Hoover and Porges (1952)
Sucrose	0·68	Placak and Ruchhoft (1947)
Dextrin	0·60	Placak and Ruchhoft (1947)
Dextrin	0·73	Placak and Ruchhoft (1947)
Ethyl alcohol	0·23	Placak and Ruchhoft (1947)
Acetic acid	0·47	Placak and Ruchhoft (1947)

strate being oxidized (Gellman and Heukelekian, 1953; Helmers, E. N. et al., 1951; Placak and Ruchhoft, 1947). Organic acids have yielded 10–60 per cent conversion to sludge while carbohydrates have yielded 65–85 per cent.

Data for various mixed wastes have shown that 25–50 per cent of the BOD removed is oxidized and the remainder synthesized to new sludge (neglecting endogenous respiration). Sawyer (1956) has shown that the expected growth of new sludge is 50–60 per cent of the dry weight of organic food. The data of Gellman and Heukelekian (1953) show a yield of 0·5 lb of volatile solids per lb BOD_5 fed to the system. Hoover et al. (1951) on dairy waste oxidation studies showed 52–58 per cent biological sludge yield from the oxidation process. Eckenfelder and Moore (1954) showed 76 per cent of the 5-day BOD removed was synthesized to new sludge in pulp and paper waste treatment. The net sludge accumulation from the oxidation of a pulp and paper mill waste in which the BOD loading to sludge ratio was maintained constant is shown in Fig. 2-14.

It can be seen from Fig. 2-1 that the sludge accumulation for an oxidation process will be the net growth (a-c) less that portion which is oxidized by the sludge mass (c-d). The net growth will depend on the starting point on the time axis which in turn depends on the BOD loading level to the process. From these considerations the resultant net accumulation of sludge from synthesis and oxidation can be expressed:

$$\Delta S(\text{lb/day VSS}) = aL_r(\text{lb/day BOD}_5 \text{ removed}) - bS_a(\text{lb/MLVSS})$$
(2-18a)

The constant a represents the fraction of 5-day BOD removed which is synthesized to new biological sludge. The absolute value of the

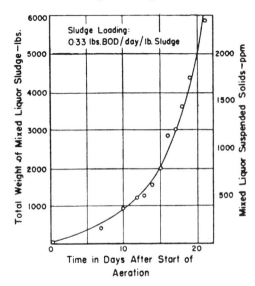

FIG. 2-14. Sludge growth at a constant BOD loading for a pulp and paper mill waste.

constant will be influenced by the relationship between the 5-day BOD and the ultimate BOD. The results of Symons and McKinney (1958) would indicate that a will increase in the presence of insufficient nitrogen. Correlation of the data of Heukelekian et al. (1951) and Wuhrman (1954) on domestic sewage oxidation showed a to be 0·49 and 0·64 respectively. The a values for rag rope waste, cotton kiering waste and brewery waste were found by Helmers et al. (1951) to be 0·49, 0·67 ad 0·54 respectively. The coefficient b is the mean rate of endogenous respiration expressed as a fraction per day.

Example 2-5.—For a given waste oxidation system the oxygen requirements are expressed:

lb O_2/day = 0·384 lb BOD_5 removed/day + 0·07 lb MLVSS

If the 5-day BOD is 70 per cent of the ultimate, derive the equation for sludge production:

Since lb of ultimate BOD removed/day

= lb O_2 consumed/day + lb VSS produced/day

and
ΔS(lb/day VSS)
$= aL_r$(lb/day BOD_5 removed) $- bS_a$(lb MLVSS)

Converting from ultimate BOD to 5-day BOD units

$$a = \frac{1 - (0·384)(0·7)}{(0·7)(1·42)} = 0·74$$

where the factor 1·42 converts from oxygen to volatile solids units and similarly

$$b' = \frac{0·07}{1·42} = 0·048$$

lb VSS_p/day

= 0·74 lb BOD_5 removed/day = 0·048 lb MLVSS.

TABLE 2-7. SLUDGE GROWTH CONSTANT
a FOR SEVERAL INDUSTRIAL WASTES

Waste	a
Spent sulfite liquor	0·55
Synthetic fibre	0·38
Pulp and paper	0·76
Refinery	0·70
Brewery	0·93
Pharmaceutical	0·77

PRINCIPLES OF BIOLOGICAL OXIDATION

Data from two waste oxidation systems are shown in Fig. 2-15. It should be noted that the auto-oxidation rate is temperature dependent. It should therefore be expected that higher net sludge accumulation will result under winter operation.

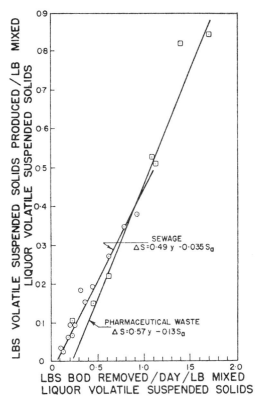

FIG. 2-15. Relationship between BOD removal and sludge accumulation for domestic sewage and a pharmaceutical waste.

In most wastes, the sludge removed from the system contains a variable portion of inert organic and inorganic solids removed from the waste waters in addition to the biological sludge. To account for this, Equation (2-18a) should be modified:

$$\Delta S(\text{lb/day VSS}) = aL_r - bS_a + I \qquad (2\text{-}18\text{b})$$

where I is the non-oxidizable volatile solids removed with the biological solids in pounds per day.

A 5-month study of a pulp and paper mill waste oxidation treating 16 Mgal/day of waste showed a gross sludge production of 12,330 lb/day for a BOD removal rate of 19,730 lb/day. The total suspended solids discharged to the aeration system from the primary sedimentation tanks was 14,000 lb/day. The total ratio of ΔS/lb BOD removed was 0·63. Correcting this value for inert suspended solids results in a ratio of ΔS/lb BOD removed of 0·49.

In an aerated lagoon, suspended solids which accumulate from synthesis and from the waste water, and which are not oxidized, settle to the bottom of the basin since the turbulence level is insufficient to maintain them in suspension. A study of an experimental lagoon treating pulp and paper mill wastes showed a net accumulation of 0·1–0·2 lb SS/lb BOD removed.

As shown in Fig. 2-1, in the absence of nutrient, biological sludge will undergo auto-oxidation resulting in a decrease in total mass. (In the analysis of bio-oxidation data it is assumed that the auto-oxidation process also occurs during sludge growth.)

The initial period of sludge oxidation will follow first-order kinetics and the rate may be conveniently expressed as a per cent per day of the sludge solids under aeration (Eckenfelder, 1956). This corresponds to a constant rate of oxidation per unit weight of sludge. After the initial oxidation, the cellular constituents remaining are progressively more difficult to oxidize and the rate declines in a logarithmic manner. After prolonged aeration, a residue will remain which is not oxidized. In the case of activated sludge from domestic sewage treatment this residue has been found to average 50 per cent. Sludge oxidation curves for several waste oxidation systems is shown in Fig. 2-16.

The rate of endogenous respiration and sludge oxidation at any time may be related to the sludge age or the mean length of time the sludge has been undergoing aeration. This sludge age is in turn related to the BOD and suspended solids loading to the process. It may be anticipated that the rate of sludge oxidation will vary inversely with sludge age. With high sludge ages, indicative of low organic loading and prolonged aeration, a large percentage of the sludge oxidation occurs in and through the aeration tanks. Further separate sludge oxidation will occur at a lower rate than that found in high rate processes. With high organic loading and low sludge ages, only a small portion of the synthesized sludge is oxidized in the aeration tanks and subsequent sludge oxidation will yield high oxidation rates. Observed oxidation rates will also vary with temperature and the nature of the waste being treated.

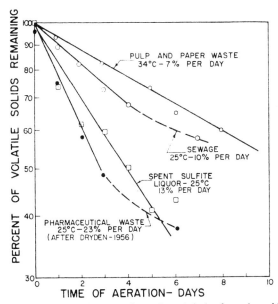

FIG. 2-16. Sludge oxidation rates for activated sludge from the oxidation of sewage and several industrial wastes.

In oxidation rate studies from a conventional activated sludge plant 10–12 per cent per day of the volatile solids were oxidized at 25°C. Over a period of six days aeration the sludge volatile content reduced from 78 per cent to 64 per cent. After 5 days aeration first-order kinetics were no longer approximated and the oxidation rate decreased rapidly with time.

The results of four oxidation studies at an activated sludge plant revealed a mean oxidation rate of 5·2 per cent per day at 21–24°C. The BOD loading to the plant during this period averaged 0·2 lb BOD/day per lb sludge. At this loading level considerable auto-oxidation occurred in the aeration tanks prior to the start of the oxidation study.

Sludge from a step-aeration plant exhibited an oxidation rate of 8 per cent per day at 19°C based on volatile solids over the first 7 days of aeration. Nitrification commenced after the third day of aeration.

When sludge from an activated sludge process treating semi-chemical and pulping wastes was aerated, 7 per cent per day of the volatile solids were destroyed by oxidation. The ratio of oxygen consumed to volatile solids destroyed was 1·57. Aerating a 1 per cent sludge Gehm (1953) found a reduction in BOD of 82 per cent after 5 days aeration.

The oxidation rate of activated sludge from pharmaceutical waste oxidation showed that a 55–60 per cent reduction in sludge solids and a 60–85 per cent reduction in BOD can be attained in a 5–6 day aeration period. The rate of volatile suspended solids reduction was approximately 25 per cent per day over the first 3 days of aeration at 25°C. The corresponding endogenous respiration rate was 14 mg O_2/hr/g volatile solids. This sludge was obtained from a pilot plant operating at a BOD loading of 2–3 lb BOD/day per lb sludge. The same oxidation rates were obtained using sludge at 2 per cent concentration (dry weight basis) as were found at a 0·4 per cent concentration indicating that concentration was not a limiting factor over this range. A high power level was maintained in the system by employing multiple turbines to insure adequate sludge dispersion.

The oxidation of sludges from dairy waste oxidation has been studied by Hoover and Porges (1952) who found a sludge oxidation rate of 24 per cent per day at 20°C.

It is apparent from Equation (2-18a) that increasing the concentration of sludge under aeration will decrease the net sludge accumulation. If the aeration process is designed such that $\Delta S \cong O$, the quantity of sludge which must be maintained under aeration becomes:

$$\frac{aL_r}{b} = S_a \qquad (2\text{-}19)$$

in which

$L_r =$ lb BOD removed/day
$S_a =$ lb volatile suspended solids under aeration

This concept has been applied to the treatment of dairy wastes in a batch fill and draw system (Kountz, 1954). An example is detailed in Chapter 6.

NUTRITIONAL REQUIREMENTS

Efficient and successful biological oxidation of organic wastes requires a minimal quantity of nitrogen and phosphorus for the synthesis of new cell tissue. In addition, trace quantities of several

INORGANIC NUTRIENT REQUIREMENTS

Element	Function
Potassium	Primarily catalytic
Calcium	Catalytic, bound to cell constituents
Phosphorus	Uncertain
Magnesium	Component of chlorophyll
Sulfur	Constituent of protein
Iron	Catalytic
Manganese, copper and zinc	Catalytic

The nitrogen cycle in biological waste treatment is:

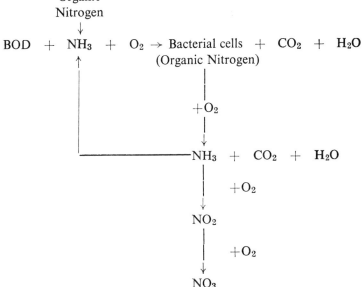

other elements such as potassium and calcium are required. These elements are usually present in natural waters in sufficient quantity to satisfy the requirements for bacterial metabolism. Nitrogen and phosphorus, however, are frequently deficient in waste substrates and must be fed as a nutrient supplement to the system to attain optimum efficiency.

During sludge growth, nitrogen is utilized for synthesis (a-c, Fig. 2-1) while during the auto-oxidation process (c-d) nitrogen is released back to solution. Some of this nitrogen will be recovered and reused for synthesis.

Nitrogen in the form of ammonia, nitrite and nitrate and some forms of organic nitrogen are available to the organisms for synthesis. The portion of organic nitrogen available varies with the waste. For example Helmers et al. (1952) found 9–23 per cent available organic nitrogen in rag rope waste and 55–78 per cent available organic nitrogen in brewery waste and domestic sewage. Soluble inorganic phosphorus and most organic phosphorus are available for microbial usage. When a nutritional supplement is required for a biological process, ammoniacal nitrogen and soluble phosphorus salts are generally used since they are most readily assimilable. It is usually not advisable to add nitrates, because they serve as a secondary source of oxygen for the organisms. In secondary settling tanks where the available dissolved oxygen may be depleted, nitrates are reduced and nitrogen gas is formed, resulting in a floating sludge. The nitrogen changes in a sewage undergoing aeration is shown in Fig. 2-17. The nitrogen changes occurring during the oxidation of a domestic sewage activated sludge is shown in Fig. 2-18.

The nitrogen of actively metabolizing sludges will vary from 6–15 per cent and phosphorus from 2–5 per cent on a dry weight basis depending on the growth phase of the system. (After initial removal of carbonaceous BOD, the nitrogen content of the cell may be low. As oxidation and sludge growth proceeds the cell nitrogen will increase to a maximum.) Symons and McKinney (1958) obtained a sludge of 8·2–8·7 per cent nitrogen from the oxidation of sodium acetate when the availability of nitrogen was not limiting. Helmers et al. (1951) reported that for optimum process efficiency a minimum nitrogen and phosphorus content of 7 per cent and 1·2 per cent by weight respectively of the total volatile solids should be maintained.

The nitrogen content of sludges is dependent on the type and nature of the active organisms produced in aerobic waste treatment systems, the concentration of biological volatile solids and the concentration of available nitrogen in the waste. The ratio of microbial mass to total volatile solids will determine the critical nitrogen

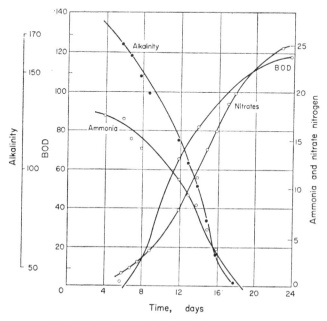

FIG. 2-17. Nitrogen relationships in the oxidation of a sewage effluent.

content. This ratio will be variable depending on the waste. For example, a biological sludge from a pulp and paper waste possessed a large percentage of stable organic matter resulting in a critical nitrogen content of 3·5 per cent based on the total volatile solids. By comparison, the nitrogen content of an average domestic biological sludge is 7·5 per cent of the total volatile solids. These solids are primarily biological in nature. It is significant that the computed value of the nitrogen based on the biological solids from the aforementioned pulp and paper waste was 7·5 per cent.

A six-month study was conducted to determine the nitrogen requirements in the oxidation of a pulp and paper mill waste. The BOD removal averaged 21,400 lb/day. Below 4·5 lb N/100 lb BOD removed, BOD removal was reduced.

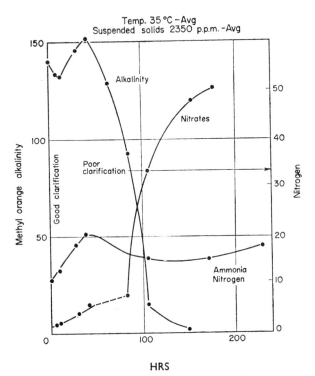

Fig. 2-18. Nitrogen relationships in the oxidation of an activated sludge.

A study of the sludge–nitrogen balance was made on the oxidation of ammonia base spent sulfite liquor. The sludge initially fed with waste liquor was in the endogenous respiration phase and had a nitrogen content of 5·45 per cent. After 24 hr aeration, synthesis had increased the nitrogen content to 13·1 per cent. After 3 days aeration in the absence of additional nutrient the nitrogen content of the sludge was reduced to 7·8 per cent.

PRINCIPLES OF BIOLOGICAL OXIDATION 65

Calculation of Nutritional Requirements

The nutritional balance of an aerobic biological system is primarily based upon satisfying the requirements of the cell structure produced by the removal of BOD from a waste. Helmers *et al.* (1951) showed that the requirements may be computed from the difference between the influent and effluent available nutrient when a minimal nutrient is present in the process effluent. This calculation requires available process data and nutrient availability data. Nutritional requirements may also be estimated from previously established values expressed as lb of nutrient required per 100 lb BOD removed from the system. Studies on cotton keiring, rag, rope and brewery wastes by Helmers *et al.* (1951) revealed that the maximum nitrogen requirement was 5–6 lb N/100 lb BOD removed and the critical requirement 3–4 lb N/100 lb BOD removed. The maximum phosphorus requirement was 1·0 lb P/100 lb BOD removed and the critical requirement 0·6 lb P/100 lb BOD removed. A BOD : N : P ratio of 100 : 5 : 1 in a waste will usually insure adequate nutrition.

Nitrogen and phosphorus requirements may be more rigorously computed from a material balance based on the maintenance of a minimum nitrogen and phosphorus content in the biological sludge produced by the system. The important variables to be considered are BOD loading, availability of nutrients, temperature, solids and time of treatment. The latter variables will influence the BOD removal efficiency of the system. The calculation of nitrogen requirement may be illustrated by the following example:

Example 2-6.—An activated sludge plant is to be designed for an industry which will produce 1·0 Mgal/day of waste. Tests on the raw waste indicate an average 5-day 20° BOD of 500 p.p.m. and a total available nitrogen content of 10 p.p.m. Removal of 90 per cent of the BOD is necessary to meet receiving stream conditions. It is assumed that approximately 60 per cent of the BOD removed is assimilated to sludge in the design aeration period,* that one pound of biological volatile solids is equivalent

* The net sludge produced in the system will vary from 30–80 per cent of the organic matter removed. This per cent will depend on the nature of the waste, the active organisms, the availability of nitrogen, and the absolute value of the endogenous respiration occurring in the system.

to 1·4 lb of ultimate oxygen demand (BOD)† and that a critical cell nitrogen content of 7 per cent based on the biological volatile solids is necessary to maintain optimum process efficiency.‡ Compute the quantity of nitrogen which must be added per day to the system.

Total nitrogen required
0·07 . 2400
170 lb per day

5-day BOD removed
1 . 8·34 . 500 . 0·90
3700 lb per day

Nitrogen available in waste
1 . 8·34 . 10
84 lb per day

Ultimate BOD removed
5500 lb per day

Net nitrogen to be fed
86 lb per day

Biological volatile solids produced

$$\frac{5500 \, . \, 0 \cdot 6}{1 \cdot 4}$$

2400 lb per day

This is equivalent to approximately 4·5 lb of nitrogen per 100 lb of 5-day BOD removed in the system.

The phosphorus requirement may be computed in a similar fashion.

Feed Control

Nitrogen may be fed to a system as a gas (anhydrous ammonia), as liquid ammonia or as a dry feed of an ammonium salt. Selection of a feed system for any given installation should be determined by a comparative cost study, considering economy, transportation, handling problems and reliability of supply. The table below shows the relative costs of nutrient chemicals for a pulp and paper waste treatment plant in the north-eastern part of the United States, based on 100 for the least costly. (In the case of ammonium phos-

† Sludges evaluated from industrial systems have been observed to require from 1·3–1·5 lb of O_2 per lb of volatile solids.

‡ Critical nitrogen content to maintain process efficiency.

phates, costs allow credit for the nitrogen content.) (Moore and Kass, 1956.)

RELATIVE COSTS OF NUTRIENT MATERIALS

Nitrogen

Anhydrous ammonia (c.l.)	100
30 per cent aqueous ammonia (c.l.)	107
Ammonium sulfate, bulk (c.l.)	170
Ammonium sulfate, bagged (l.c.l.)	250
Anhydrous ammonia, 150 lb cylinders	300

Phosphorus

75 per cent phosphoric acid, bulk (c.l.)	100
75 per cent phosphoric acid, drums (l.c.l.)	133
Monoammonium phosphate (l.c.l.)	126
Diammonium phosphate (l.c.l.)	133
Monosodium phosphate (l.c.l.)	165
Trisodium phosphate (l.c.l.)	256

EFFECT OF TEMPERATURE

Temperature influences the rate of all chemical and biochemical reactions. In most reactions occurring in the range of optimum biological activity, a two to three fold increase in reaction velocity is experienced for each 10° rise in temperature. The variation in k_r can be expressed by the Van't Hoff–Arrhenius equations:

$$\frac{1}{K}\frac{dK}{dT} = \frac{\Delta E}{RT^2}$$

where

$R =$ Universal gas constant
$T =$ Absolute temperature
$C =$ Constant
$K =$ Reaction rate
$\Delta E =$ const. $=$ energy of activation

integrating

$$\ln K = \frac{-\Delta E}{RT} + C$$

or
$$\log k = \frac{-\Delta E}{2 \cdot 3\ R} \cdot \frac{1}{T} + C$$

or
$$\log \frac{k_2}{k_1} = \frac{\Delta E}{2 \cdot 3\ R} \left[\frac{T_2 - T_1}{T_2 \cdot T_1}\right]$$

from a plot of $\log_{10} k$ vs. $1/T$

$$\text{slope} = \frac{-\Delta E}{2 \cdot 3\ R} = \frac{-\Delta E}{4 \cdot 58}$$

ΔE (activated sludge) = 14,400 cal/mole

ΔE (biological reactions) = 8000–18,000 cal/mole

Formulae for Temperature Effect

(1) $\log k_2/k_1 = 0 \cdot 0368(t_1 - t_2)$; 0–28°C   (Fig. 2-19)

(2) $0 \cdot 0315 = \dfrac{\log k_1 - \log k_2}{t_1 - t_2}$ Wuhrman (1954)

(3) $\dfrac{k_t}{k_{25°c}} \times 100 = 0 \cdot 71 t^{1 \cdot 54}$ Sawyer and Rohlich (1939)

(4) $k_T = k_{20°c} \cdot 1 \cdot 065^{(t-20)}$ Phelps (1944)

The temperature influence on bio-oxidation processes has been observed to be less than that estimated from the foregoing equations. Since all processes represent a sequence of related reactions (oxygen transfer, absorption, oxidation, etc.) it is probable that temperature effects on oxidation are not the controlling reaction under all temperature conditions. For example, at high temperatures, oxygen transfer may control while at low temperatures the respiration may control. Since most conventional plants function at low loading levels it is probable critical loading levels are never exceeded even at low operating temperatures.

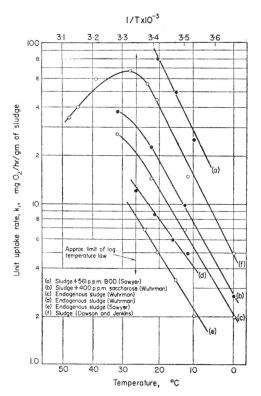

Fig. 2-19. Effect of temperature on the oxygen utilization rate of activated sludge under various loading conditions.

EFFECT OF pH

The effect of pH on the overall oxidation process is that normally associated with specific enzymatic processes. Over some pH range for each particular enzyme the activity approaches a maximum and falls off above or below this range. In a heterogeneous system as encountered in bio-oxidation processes involving a whole sequence of enzyme reactions a mean pH range will be established as shown in Fig. 2-20.

In many waste oxidation systems, the pH will tend to approach pH 8·0. In alkaline wastes, reaction between the CO_2 produced from respiration and the carbonate and hydroxide will form from bicarbonate while in wastes containing organic acids the CO_2 from the oxidation of the acids is scrubbed out by aeration resulting in a pH rise. The oxidation of salts of organic acids produce basic anhydrides which in reaction with CO_2 form bicarbonate. The bicarbonate will

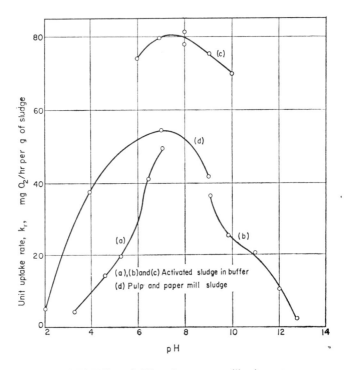

FIG. 2-20. Effect of pH on the oxygen utilization rate of activated sludge.

buffer the system about pH 8·0. Typical relationships obtained from the oxidation of a pulp and paper mill waste is shown in Fig. 2-21. Neutralization and pH adjustment will usually be required in the presence of mineral acids or caustics. Sawyer *et al.* (1955) has indicated the desirability of pH adjustment in those systems where weakly ionized acids are converted to highly ionized acids (sulfite liquor) and where neutral substances are converted to acidic forms by oxidation (formaldehyde). The deleterious effects of pH are magnified at low temperature.

The effective pH range for cotton kiering liquor and sulfite waste is pH 5–11; for spent yeast broth, slaughterhouse waste and candied fruit waste pH 5–9; for boardmill white water and rag cook liquor pH 6–9; and antibiotics pH 5–7: (Sawyer *et al.*, 1955).

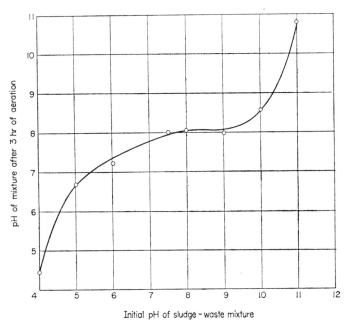

FIG. 2-21. pH relationships during biological oxidation.

Keefer and Meisel (1951) found the optimum pH range for activated sludge treating domestic sewage to be pH 7·0–7·5 with an effective process range of 6·0–9·0. At pH 4·0 the process was only 43 per cent effective and at pH 10·0 only 54 per cent effective. A rapid change in pH (for example a drop in pH from 6 to 5) may decrease the respiratory activity by as much as 75 per cent. In many systems the optimum zone for biological activity may vary appreciably from neutrality. pH effects in two typical bio-oxidation systems are summarized below.

Case (1)—alkaline wastes

$$\text{Waste} + (OH^-) \qquad pH > 8.6$$
$$\downarrow O_2, CO_2$$
$$\downarrow \begin{array}{l} +OH^- \\ +CO_3^- \end{array}$$
$$HCO_3^- \qquad pH\ 8\text{–}8.3$$

Case (2)—organic acids

$$HAc,\ NH_4Ac \qquad pH\ 3\text{–}6$$
$$\downarrow O_2$$
$$CO_2 + H_2O + NH_4^+ \text{ (excess } CO_2 \text{ scrubbed out)}$$
$$NH_4HCO_3;\ NH_{4_2}CO_3 \qquad pH\ 8\text{–}9$$
$$\downarrow O_2$$
$$NO_3,\ CO_2,\ H_2O$$
$$NH_4NO_3,\ NaNO_3 \qquad pH\ 6\text{–}7$$

Toxicity

Many substances exert a toxic effect on biological oxidation processes. Partial or complete inhibition may result, depending on the substance and its concentration. Inhibition may result from interference with the osmotic balance or with the enzyme system. The concentration of a substance which will exert a toxic effect is also influenced by other factors such as food concentration, temperature, and the nature of the organisms (Ingols, 1955). In some cases the biological population will acclimatize itself to a concentration level of a toxic substance. This acclimatization may result from a neutralization of the toxic material by the biological activity of the microorganisms or a selective growth of the culture in which only those organisms capable of metabolism in the presence of the toxic substance will proliferate. In some cases such as cyanide, the toxic material may be used as a food substance. A detailed study of toxicity in phenol oxidation was made by Abson et al. (1959). They showed that ammonium in concentrations in excess of 1700 p.p.m. inhibited the rate of phenol oxidation. Inhibition was complete at concentrations in excess of 5000 p.p.m. It is believed that inhibition results from an increase in osmotic pressure at high ammonium ion concentration. In the presence of non-acclimatized sludge, sulfide completely inhibited biological growth in concentrations in excess of 25 p.p.m. Adaption for 24–48 hr increased the tolerance level of sulfide to 100 p.p.m. In the case of heavy metals such as Copper, Zinc and Chromium, adaption for 72 hr increased the tolerance level from 1–5 p.p.m. to greater than 75 p.p.m.

REFERENCES

1. ABSON, J. W. and TODHUNTER, K. H. "Factors Affecting the Biological Treatment of Carbonization Effluents": paper presented at Cake Oven Managers Assoc., (Feb. 1959).
2. BOGAN, R. H. and SAWYER, C. N., *Sew. and Ind. Wastes*, **27**, 8, 917 (1955).
3. BUSCH, A. W. and KALINSKE, A. A. *Biological Treatment of Sewage and Industrial Wastes* Vol. I. (Ed. by MCCABE, B. J. and ECKENFELDER, W. W.) Reinhold Pub. Corp., New York, N.Y. (1956).
4. BUSWELL, A., VAN MELER, J. and GERKE, J. R., *Sew. and Ind. Wastes*, **22**, 4, 508 (1950).
5. DAWSON, P. S. S. and JENKINS, S. H., *Sew. Wks. J.*, **21**, 4, 643 (1949).
6. DAWSON, P. S. S. and JENKINS, S. H., *Sew. and Ind. Wastes*, **22**, 4, 490 (1950).
7. ECKENFELDER, W. W. and MOORE, T. L., *Eng. News Record*. (Dec. 6, 1954).
8. ECKENFELDER, W. W. and O'CONNOR, D. J., Proc. 9th Ind. Waste Conf. Purdue Univ. (1954).
9. ECKENFELDER, W. W., *Sew. and Ind. Wastes*, **28**, 8 (1956).
10. ECKENFELDER, W. W. and HOOD, J. W., *Water and Sewage Wks.*, (June 1950).
11. ECKENFELDER, W. W., Proc., 14th Ind. Waste Conf., Purdue Univ. (1959).
12. ECKENFELDER, W. W. and MCCABE, B. J., *Waste Treatment* (Ed. by ISAAC, P.) Pergamon Press, Oxford (1960).
13. FAIR, G. M. and MOORE, E. W., *Sew. Wks. J.*, **4**, 429 (1932).
14. FAIR, G. M. and MOORE, E. W., *Eng. News Record*, **114**, 681 (1935).
15. FAIR, G. M. and GEYER, J. C. *Water Supply and Waste Water Disposal*. John Wiley and Sons, Inc., New York (1954).
16. GADEN, E. *Biological Treatment of Sewage and Industrial Wastes* Vol. I, (Ed. by MCCABE, B. J. and ECKENFELDER, W. W.) Reinhold Pub. Corp., New York, N.Y. (1956).
17. GAMESON, A. L. H. and WHEATLAND, A. B., *J. and Proc. Inst. of Sew. Purif.* Part 2 (1958).

18. GARRETT, T. M. and SAWYER, C. N., *Proc. 7th Ind. Waste Conf.*, *Purdue Univ.* 51 (1952).
19. GEHM, H. p. 346 *Proc. 8th Ind. Waste Conf.*, *Purdue Univ.* (1953).
20. GELLMAN, I. and HEUKELEKIAN, H., *Sew. & Ind. Wastes*, **22**, 10, 1321 (1950).
21. GELLMAN, I. and HEUKELEKIAN, H., *Sew. & Ind. Wastes*, **27**, (70) 793 (1955).
22. GELLMAN, I. and HEUKELEKIAN, H., *Sew. and Ind. Wastes*, **25**, 10, 1196 (1953).
23. GRANT, S., HURWITZ, E. and MOHLMAN, F. W., *Sew. Wks. J.*, **2**, 2, 228 (1930).
24. GREENHALGH, R. E., JOHNSON, R. L. and NOTT, H. D., *Chem. Eng. Prog.*, **55**, 44 (1959).
25. HAWKES, H. A., *Waste Treatment* (Ed. by ISAAC P.) Pergamon Press, Oxford (1960).
26. HELMERS, E. N., FRAME, J. D., GREENBERG, A. F. and SAWYER, C. N. *Sew. and Ind. Wastes*, **23**, 7, 834 (1951).
27. HELMERS, E. N., FRAME, J. D., GREENBERG, A. F. and SAWYER, C. N. *Sew. and Ind. Wastes*, **24**, 4, 496 (1952).
28. HEUKELEKIAN, H., ORFORD, H. E. and MANGANELLI, R., *Sew. and Ind. Wastes*, **23**, 7, 945 (1951).
29. HEUKELEKIAN, *Sew. Wks. J.*, **19**, 5, 875 (1947).
30. HIXON, A. W. and GADEN, E. L., *Ind. Eng. Chem.* **42**, p. 1792 (1950).
31. HOBER, HITCHCOCK et al., *Physical Chemistry of Cells and Tissues* Blakiston Pub. Co., New York (1950)
32. HOOVER, S. R., JASEWICZ, L., PEPINSKY, J. B. and PORGES, N., *Sew. & Ind. Wastes J.*, **23**, 2, 167 (1951).
33. HOOVER, S. R., JASEWICZ, L. and PORGES, N., *Sew. and Ind. Wastes J.*, **24**, 9, 1144 (1952).
34. HOOVER, S. R. and PORGES, N., *Sew. & Ind. Wastes*, **24**, 3, 306 (1952).
35. HOOVER, S. R., JASEWICZ, L. and PORGES, N., *Sew. and Ind. Wastes*, **25**, 10, 1163 (1953).
36. HOOVER, S. R., JASEWICZ, L. and PORGES, N., *Water and Sewage Wks.*, (1954).
37. INGOLS, R., *Sew. and Ind. Wastes*, **27**, 12, 26 (Jan. 1955).
38. JASEWICZ, L. and PORGES, N., *Sew. and Ind. Wastes*, **28**, 9, 1130 (1956).
39. KEEFER, C. E. and MEISEL, J., *Sew. and Ind. Wastes*, **27**, 3, 982 (1951).
40. KOUNTZ, R. R., *Food Eng.*, **89**, 90 (1954).
41. KOUNTZ, R. R. and FORNEY, C., *Sew. and Ind. Wastes*, **31**, 7, 810 (1959).
42. MCKINNEY, R. E. and HORWOOD, M. P., *Sew. and Ind. Wastes*, **24**, 2, 117 (1952).
43. MCKINNEY, R. E., *Sew. and Ind. Wastes*, **24**, 3, 280 (1952).
44. MCKINNEY, R. E. and JERIS, J. S., *Sew. and Ind. Wastes*, **27**, 6, 728 (1955).
45. MCKINNEY, R. E., TOMLINSON, H. D. and WILCOX, R. L., *Sew. and Ind. Wastes*, **28**, 4, 547 (1956).
46. MCKINNEY, R. E., and ENGLEBRECT, R. S., *Sew. and Ind. Wastes*, **29**, 1350–62 (1957).
47. MCKINNEY, R. E. *Biological Treatment of Sewage and Industrial Wastes* Vol. I, (Ed. by MCCABE, B. J. and ECKENFELDER, W. W.) Reinhold Pub. Corp., New York, N.Y. (1956).
48. MCKINNEY, R. E., Proc. Third Biological Waste Treatment Conference, Manhattan College (1960).
49. MOORE, E. E., *Sew. Wks. J.*, **9**, 1, 12 (Jan. 1937).

50. MOORE, T. L. and KASS, E. A., *Biological Treatment of Sewage and Industrial Wastes*, Vol. I (Ed. by MCCABE, B. J. and ECKENFELDER, W. W.) Reinhold Pub. Corp., New York, N.Y. (1956).
51. NELSON, D. J., *Sew & Ind. Wastes*, 26, 9, 1126 (1954).
52. PASVEER, A., *Sew. & Ind. Wastes*, 26, 1, 28 (1954).
53. PHELPS, E. B. *Stream Sanitation* John Wiley & Sons, New York (1944).
54. PLACAK, O. R. and RUCHHOFT, C. C., *Sewage Works J.*, 19, 3, 423 (1947).
55. PORGES, N., JASEWICZ, L. and HOOVER, S. R., *Sew. & Ind. Wastes*, 24, 9, 1091 (1952).
56. PORGES, N., JASEWICZ, L. and HOOVER, S. R., *Proc. 10th Ind. Waste Conf.*, Purdue Univ. (1955).
57. PORGES, N., *J. Chem. Educ.*, 30, 562 (1953).
58. SAWYER, C. N. and NICHOLS, M., *Sew. Wks. J.*, 11, 3, 462 (1939).
59. SAWYER, C. N. and ROHLICH, G. A., *Sew. Wks. J.*, 11, 6, 946 (1939).
60. SAWYER, C. N., *Sew. & Ind. Wastes*, 27, 8, 929 (1955).
61. SAWYER, C. N., *Biological Treatment of Sewage and Industrial Wastes*, Vol. I, (Ed. by MCCABE, B. J. and ECKENFELDER, W. W.) Reinhold Pub. Corp., New York, N.Y. (1956).
62. SAWYER, C. N., FRAME, J. D. and WOLD, J. P., *Sew and Ind. Wastes* 27, 929 (1955).
63. STACK, V. T., *Sew. & Ind. Wastes*, 29, 9, 987 (1957).
64. STERN, A., West Virginia Pulp and Paper Co., Covington, Va. Technical Report (1959).
65. STREETER, H. W., *Sew. Works J.*, 6, 2, 208 (1934).
66. SYMONS, J. M. and MCKINNEY, R. E., *Sew. & Ind. Wastes*, 30, 7, 874 (1958).
67. TAMIYA, H. "Material and Energy Balances of Biological Synthesis" Actualities Scientifiques et Industrielles, 214, Exposes de Biologie (1935).
68. THERIAULT, E. J. and MCNAMEE, P. H., *Ind. Eng. Chem.*, 22, 12, 1330 (1930).
69. VELZ, C. J., *Sew. Works J.*, 20, 4, 607 (1948).
70. Water Pollution Research, Dept. of Scientific and Industrial Research (Brit.) (1956).
71. WESTON, R. F. and ECKENFELDER, W. W., *Sew. & Ind. Wastes*, 27, 7, 802 (1955).
72. WESTON, R. F., STACK, V. and SITMAN, W., "Mass Transfer Relationships in Biological Waste Treatment", Pres. A.I.Ch.E. Annual Meeting, Atlanta, Ga. (Feb. 1960).
73. WUHRMAN, K., *Sew. & Ind. Wastes*, 26, 1, 1 (1954).
74. WUHRMAN, K., *Biological Treatment of Sewage and Industrial Wastes* Vol. I, (Ed. by MCCABE, B. J. and ECKENFELDER, W. W.) Reinhold Pub. Corp. (1956).
75. WUHRMANN, K., *Schweiz. Zeits. Hydrol.* 20, 284–330 (1958).
76. WUHRMAN, K. *Proc. Third Biological Waste Treatment Conference*, Manhattan College (1960).

CHAPTER 3

THEORY AND PRACTICE OF AERATION

OXYGEN SATURATION

Oxygen is a sparingly soluble gas in pure water, having a saturation at 20°C of 9.02p.p.m. in equilibrium with the atmosphere. While the solubility is almost independent of the total pressure and the presence of other gases, it is directly proportional to the partial pressure of oxygen in the gas phase. The relationship is defined by Henry's law which states: When no chemical reaction is involved, the quantity of gas which will dissolve at a given temperature is directly proportional to the partial pressure of the gas in contact with water,

$$C_s = H_s p \qquad (3\text{-}1)$$

in which C_s is the oxygen saturation, p the partial pressure of oxygen in the gas and H_s Henry's law constant.

By raising the partial pressure of oxygen in the gas phase, the driving force for transfer from gas to liquid can be increased. The factors of importance which influence the constant, H_s, are the temperature and the presence of dissolved and other solids.

Gases exhibit decreasing solubility with increasing temperature. The saturation of oxygen in water at various temperatures may be computed from the equation:

$$C_s = 14.16 - 0.3943\,T + 0.007714\,T^2 - 0.0000646\,T^3 \qquad (3\text{-}2)$$

where T is in °C.

A simpler equation which is sufficiently accurate for most purposes has been proposed by Gameson and Robertson (1955)

$$C_s = \frac{475 - 2.65\,S}{33.5 + T} \qquad (3\text{-}3)$$

The saturation values obtained from these equations must be multiplied by the ratio of the prevailing barometric pressure to the standard pressure of 760 mm Hg. The term 2·65 S provides a salinity correction.

Since air is usually released at a 12–15 ft depth in aeration tanks, its partial pressure is increased by the liquid submergence. As the air bubbles rise through the tank, oxygen is absorbed, reducing the oxygen concentration in the gas phase. The effect of these factors must be considered in deriving a mean saturation value, as follows (Oldshue, 1956):

$$C_s = C_w \left(\frac{P_b}{29 \cdot 4} + \frac{O_t}{42} \right) \qquad (3\text{-}4)$$

C_s is the oxygen saturation value corresponding to the average partial pressure of oxygen in the gas stream entering and leaving the aerator. C_w is the oxygen saturation value in the waste at atmospheric pressure. P_b is the absolute pressure in pounds per square inch at the depth of air release. O_t is the percent concentration of oxygen in the air leaving the tank. The effect of water vapor pressure is neglected in Equation (3-4). While the mean saturation value for rising air bubbles is defined by Equation (3-4), that for surface aeration will be defined by C_s at atmospheric conditions. The saturation value during bubble formation is defined by the partial pressure at the point of air release. The value computed for the mid-depth of the tank is a representative average for most applications.

The presence of dissolved solids influence the value of oxygen saturation in water. The solubility of oxygen in sea water of various concentrations has been reported by Gameson and Robertson (1955). Five thousand p.p.m. of chlorides reduce the solubility of oxygen 5 per cent. It is probable that many waste substances have similar effects on the saturation value. Oxygen saturation in sewage is approximately 95 per cent that of pure water. Various waste streams from pulp and paper manufacture have shown variations of 85–95 per cent of pure water saturation.

Example 3-1. Sewage is being aerated at 24°C. The air diffusers are located 15 ft below the water level in the basin. The estimated oxygen absorption is 10 per cent. Compute the mean saturation value in the tank.

C_W for sewage at 24°C $= 8.5 \times 0.95 = 8.07$ p.p.m.

Moles of O_2 in exit gas $= 21 \times 0.9 = 18.9$

Moles of N_2 in exit gas $= 79.0$

Per cent O_2 in exit gas $= \dfrac{18.9}{18.9 + 79.0} \times 100$ per cent $= 19.3$ per cent

P.s.i.a. at tank bottom $= 14.7 + 6.5 = 21.2$ p.s.i.a.

$C_s \qquad\qquad = 8.07 \left(\dfrac{21.2}{29.4} + \dfrac{19.3}{42}\right)$

$C_s \qquad\qquad = 8.07 \,(0.72 + 0.46) = 9.53$ p.p.m.

THEORY OF OXYGEN TRANSFER

Oxygen disperses itself through a body of liquid by the process of diffusion which tends to produce a stable state of uniform concentration. Mass transfer by diffusion occurs between two phases when a driving force is created by a departure from equilibrium. This driving force is a partial pressure gradient in the gas phase and a concentration gradient in the liquid phase. The rate of molecular diffusion of a dissolved gas in a liquid is dependent on the characteristics of the gas and the liquid, the temperature, the concentration gradient and the cross-sectional area through which diffusion occurs. This condition is shown diagrammatically in Fig. 3-1.

The diffusional process is defined by Fick's law:

$$\frac{dm}{dt} = - D_L A \frac{dc}{dy} \qquad (3\text{-}5)$$

where dm/dt is the time rate of mass transfer by diffusion; A is the cross-sectional area through which diffusion occurs, dc/dy is the concentration gradient (mass/unit volume/unit length) in the direction perpendicular to the cross-sectional area, and D_L is the diffusion coefficient for oxygen in water. D_L is expressed as area per unit time. Mass transfer occurs through laminar films at the gas and liquid

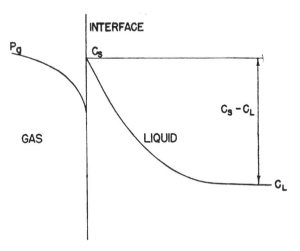

FIG. 3-1. Schematic representation of the oxygen transfer mechanism.

interfaces and through the turbulent body of fluid until a dynamic equilibrium is established. This transfer from the atmosphere to a body of turbulent fluid may be described by the Lewis and Whitman (1924) two film concept expressed in differential form similar to Fick's law of hydro-diffusion. This leads to the equation:

$$\frac{dm}{dt} = - D_g A \left[\frac{\partial c}{\partial y}\right]_1 = - D_L A \left[\frac{\partial c}{\partial y}\right]_2 = - D_e A \left[\frac{\partial c}{\partial y}\right]_3 \quad (3\text{-}6)$$

in which,

$\left[\dfrac{\partial c}{\partial y}\right]_1$ = concentration gradient through the gas film

$\left[\dfrac{\partial c}{\partial y}\right]_2$ = concentration gradient through the liquid film

$\left[\dfrac{\partial c}{\partial y}\right]_3$ = concentration gradient in the body of the liquid below the liquid film

D_g = molecular diffusivity of the gas through the gas film

D_L = molecular diffusivity of the gas through the liquid film

D_e = eddy diffusion coefficient of the gas in the body of the liquid

In addition to the molecular diffusion through the gas and liquid films Equation (3-6) also includes the eddy diffusion through the body of the liquid. In the vast majority of cases of aeration, turbulent flow prevails. The value of the eddy diffusion coefficient is the same order of magnitude as that of the eddy viscosity of the fluid, which depends upon the physical and hydraulic characteristics of the system. Since the eddy diffusivity is so great by comparison to the diffusivity through the liquid film, it follows from Equation (3-6) that the concentration gradient throughout the depth of the liquid is extremely small by comparison to the gradient across the liquid film. In all practical cases, then, the concentration of dissolved oxygen throughout the depth of the liquid body may be taken as uniform. This has been substantiated by actual measurements of dissolved oxygen gradients. The magnitude of the diffusivity in the gas phase is 10^4 greater than that in the liquid phase. It may therefore be concluded from these considerations that in the transfer of a sparingly soluble gas such as oxygen, the controlling resistance is in the liquid film.

Since the concentration of dissolved oxygen may be taken as uniform throughout the depth, the entire gradient may be assumed to exist at the interface and Equation (3-6) may be re-expressed as follows:

$$\frac{dm}{dt} = \frac{D_L}{y_L} A (C_s - C) = K_L A (C_s - C) \qquad (3\text{-}7)$$

The liquid film coefficient K_L is defined as the diffusivity divided by the hypothetical film thickness, y_L, Equation (3-7) may be ex-

THEORY AND PRACTICE OF AERATION

pressed in concentration units by introducing the volume of the liquid:

$$\frac{1}{V} \cdot \frac{dm}{dt} = \frac{dc}{dt} = K_L \frac{A}{V}(C_s - C) \tag{3-8}$$

Equation (3-8) shows that the rate of transfer is proportional to the concentration gradient, the ratio of the interfacial area to the liquid volume and the transfer coefficient K_L. It is important to note that the ratio of area to volume is readily defined for natural streams,

$$\frac{A}{V} = \frac{1}{H}$$

but difficult to determine in artificial aeration systems. Due to the difficulty of measuring interfacial areas, an overall transfer coefficient, $K_L a$ is usually employed.

$$\frac{dc}{dt} = K_L a (C_s - C) \tag{3-9}$$

$K_L a$ is a function of the interfacial area, volume of liquid, and other physical and chemical variables characteristic of the system. Since $C_s - C =$ dissolved oxygen deficit, D, Equation (3-9) may also be expressed as follows:

$$\frac{dD}{dt} = -K_2 D \tag{3-10}$$

which integrates to

$$D = D_0 \, 10^{-k_2 t} = D_0 \, e^{-k_2 t} \tag{3-11}$$

in which

$K_2 = K_L a = 2 \cdot 30 \, k_2 =$ reaeration coefficient

$D =$ dissolved oxygen deficit after time, t

$D_0 =$ initial dissolved oxygen deficit

Danckwertz (1951) defined the liquid film coefficient as the square root of the product of the diffusivity and the rate of surface renewal:

$$K_L = (D_L r)^{1/2} \tag{3-12}$$

The rate of surface renewal, r, may be considered as the frequency with which fluid with a solute concentration C is replacing fluid from the interface with a concentration C_s. This establishes a layer with an equilibrium concentration C_s at the interface and a concentration C in the bulk of solution. The thickness of this layer (analogous to film thickness) will vary with fluid turbulence. Similar definitions of the liquid film coefficient have been developed by Higbie (1935).

STREAM AERATION

The rate of surface renewal, r, has been defined by O'Connor and Dobbins (1958), by fluid turbulence parameters, viz., the mixing length and the vertical velocity fluctuation as follows:

$$r = \frac{\bar{v}}{\bar{l}} \tag{3-13}$$

In the case of reaeration in natural waters, these parameters have been related to the physical characteristics of the channel and flow. Two cases are considered:

1. Non-isotropic turbulence where a velocity gradient and shearing stress exist, which is characteristic of shallow streams. The ratio of the vertical velocity fluctuation and mixing length is equal to the velocity gradient at the surface and the liquid film coefficient is

$$K_L = \left(D_L \frac{dU}{dy} \right)^{1/2} \tag{3-14}$$

It has been shown that under these conditions the reaeration coefficient may be expressed as follows:

$$k_2 = \frac{480 \, D_L^{1/2} s^{1/4}}{H^{5/4}} \tag{3-15}$$

in which,

D_L = coefficient of molecular diffusion, ft² per day

s = slope of the river channel, ft per ft

H = average depth of the stream, ft

k_2 = reaeration coefficient, per day

2. Isotropic turbulence where neither a significant velocity gradient nor shearing stress exist, which is characteristic of deep streams. In this case the vertical velocity fluctuation and the mixing length are approximately equal to one-tenth of the forward flow velocity and the average depth respectively. The liquid film coefficient:

$$K_L = \left(\frac{D_L U}{H}\right)^{1/2} \qquad (3\text{-}16)$$

The reaeration coefficient is expressed as follows:

$$k_2 = \frac{(D_L U)^{1/2}}{2 \cdot 3 \, H^{3/2}} \qquad (3\text{-}17)$$

in which,

U = the average velocity of the stream and the other terms as previously defined.

The distinction between the two types of turbulence is established by the roughness coefficient of the channel, defined as follows:

$$U = B\sqrt{Hs} \qquad (3\text{-}18)$$

in which,

B = Chezy coefficient

If the value of B is less than 17, the turbulence is considered as non-isotropic and if B is greater than 17, turbulence is considered as isotropic. It is significant to note that Equation (3-17) may be also used as an approximation to define the reaeration coefficient for non-isotropic turbulence. A comparison of the observed coefficients and those calculated by means of Equation (3-17) is shown in Fig. 3-2.

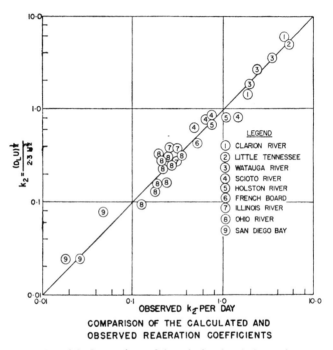

Fig. 3-2. Comparison of the calculated and observed reaeration coefficients.

Example 3-2. A river survey has indicated the following for a drought flow condition:

Length of stretch	2·5 miles
average depth	1·2 ft
average velocity	0·75 ft/sec
slope	0·0021 ft/ft
temperature	20°C

Determine the reaeration coefficient.

$$B = U/\sqrt{Hs}$$

$$= 0.75/\sqrt{1.2 \times 0.0021}$$

$$= 15$$

The turbulence is non-isotropic and Equation (3-15) applies.

$$k_2 = \frac{480\ D_L^{1/2}\ s^{1/4}}{H^{5/4}}$$

For 20°C,

$$D_L = 0.00195 \text{ ft}^2/\text{day}$$

$$k_2 = \frac{480\ (0.00195)^{1/2}\ (0.0021)^{1/4}}{1.2^{5/4}}$$

$$= 3.6/\text{day}$$

At a higher river flow at 20°C the following condition was observed:

average depth	2.6 ft
average velocity	2.1 ft/sec
average slope	0.0021 ft/ft

$$B = 2.1/\sqrt{2.6 \cdot 0.0021}$$

$$= 28$$

The turbulence is isotropic and Equation (3-17) applies.

$$k_2 = \frac{(D_L U)^{1/2}}{2 \cdot 3 \, H^{3/2}}$$

$D_L = 0.0000812$ ft²/hr and $U = 7550$ ft/hr

$$k_2 = \frac{(24 \; 0.0000812 \; . \; 7550)^{1/2}}{2 \cdot 30 \; . \; 2 \cdot 6^{3/2}}$$

$$= 1 \cdot 52/\text{day}$$

BUBBLE AERATION

In diffused aeration, air bubbles are formed at an orifice, from which they break off and rise through the liquid, finally bursting at the liquid surface. Oxygen transfer occurs as the bubble emerges from the orifice, as the bubble rises through the liquid, and as they burst at the surface shedding an oxygen saturated film into the surface layers. Additional surface reaeration occurs from velocity gradients induced at the surface from turbulence generated by the rising bubbles. Since the liquid film coefficient is higher during bubble formation and release than during bubble rise through the liquid it may be expected that K_L will vary inversely with liquid depth.

Bubble Velocity and Size

The velocity and shape characteristics of air bubbles in water can be related to a modified Reynolds number (Hoberman and Morton, 1956). At *Re* less than approximately 300, the bubbles are spherical and act as rigid spheres. The bubble rise is rectilinear or helical. Over a *Re* range of 300–4000 the bubbles assume an ellipsoidal shape and rise with a rectilinear, rocking motion. The bubbles form spherical caps at *Re* greater than 4000.

The rising velocity of the bubbles is increased at high air flows due to the proximity of other bubbles and resulting disturbances of the bubble wakes.

The size of air bubbles released by diffused aeration devices is related to both the orifice diameter and the air rate. At low air rates the bubble volume is directly proportional to the orifice diameter and the surface tension and inversely proportional to the liquid density. The bubble size produced will result from a balance of the buoyant force separating the bubble from the orifice and the shearing force necessary to break the surface tension across the orifice (Ippen and Carver, 1955). The bubble size is independent of air rate and the frequency of bubble release is directly proportional to air rate. At high air rates the bubble diameter increases as a function of the gas flow rate, and the frequency is constant. An intermediate zone exists during which both frequency and bubble volume are changing. Over the range of air rates normally encountered in aeration practice, the mean diameter of bubble produced is an exponential function of the gas rate:

$$d_B \sim G_s^n \qquad (3\text{-}19)$$

in which

d_B = mean bubble diameter

G_s = air flow

n = exponent

Correlation of Bubble Aeration Data

Eckenfelder (1959) showed that for any aeration depth, the oxygen transfer characteristics could be correlated according to the dimensionless Sherwood, Reynolds and Schmidt numbers:

$$\frac{K_L d_B}{D_L} = F\left(\frac{d_B v_B}{\nu}\right)\left(\frac{\nu}{D_L}\right)^{1/2} \qquad (3\text{-}20a)$$

in which

v_B = bubble velocity

ν = kinematic viscosity

For aeration depths greater than approximately 100 cm Equation

(3-20a) can be generalized for all depths by applying an exponential depth correction to compensate for the end effects:

$$\frac{K_L d_B}{D_L} \cdot H^{1/3} = F\left(\frac{d_B v_B}{\nu}\right)\left(\frac{\nu}{D_L}\right)^{1/2} \qquad (3\text{-}20b)$$

Bubble aeration data is correlated according to Equation (3-20b) in Fig. (3-3). End effects may also be corrected for by correlating the

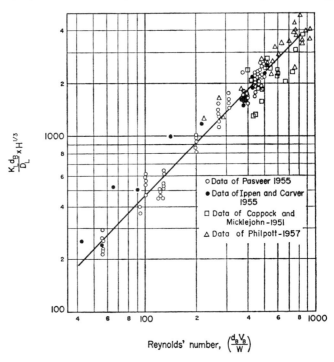

FIG. 3-3. Correlation of variables for oxygen transfer from air bubbles in water.

coefficient, F, in Equation (3-20a) against depth and extrapolating the correlation to zero depth.

From laboratory data, for a 2·4 mm diameter air bubble rising through water at 15°C Scouller and Watson (1934) determined K_L as 140 cm/hr. Adeney and Becker (1920) found variations from

32 to 230 cm/hr depending on the surface location of the bubble (tip or side). Ippen *et al.* (1952) reported values of K_L of 110–200 cm/hr at various submergence depths.

When evaluating the performance of commercial aeration equipment it is convenient to consider the aeration process in terms of the overall mass transfer coefficient $K_L a$. In bubble aeration processes, the interfacial area volume ratio can be determined.

The number of bubbles generated per minute is:

$$\frac{\text{gas flow at the orifice}}{\text{volume per bubble at the orifice}} = \frac{G_s}{(\pi/6)\, d_B^3}$$

The total surface area, A, per minute is:

$$(\text{no. bubbles/min})(\text{area/bubble}) = \frac{G_s}{(\pi/6)\, d_B^3} \cdot \pi d_B^2 = \frac{6\, G_s}{d_B}$$

The contact time of the bubble in the aeration tank is:

$$\frac{\text{tank depth}}{\text{bubble velocity}} = \frac{H}{v_B}$$

The total surface area in the tank at any time is therefore:

$$\frac{6\, G_s}{d_B} \cdot \frac{H}{v_B}$$

and the ratio A/V is:

$$\frac{6\, G_s H}{d_B v_B V} \qquad (3\text{-}21)$$

Since oxygen transfer is also contributed from the turbulent surface, Equation (3-21) becomes:

$$\frac{6\, G_s H}{d_B v_B V} + F'\left(\frac{1}{H}\right) \qquad (3\text{-}22)$$

For the depths normally employed in aeration practice (10–15 ft) the term $1/H$ becomes small with respect to the bubble surface and can usually be neglected. For example, in an aeration tank of 15 ft depth with an air flow of 20 scfm/1000 ft³ and a bubble diameter of 0·1 in., A/V for the bubbles will be 1·8 and for the surface 0·067. $1/H$ would represent a smooth surface, while actually the interfacial area will be considerably greater owing to the high degree of surface turbulence. To compensate for this a constant F' is included in Equation (3-22). F' is greater than 1·0.

Downing (1960) showed that the contribution of oxygen transfer from surface aeration varied from 4·45 per cent at a 4 ft liquid depth to 2·21 per cent at a 12 ft liquid depth.

Combining Equations (3-21) and (3-20b) and neglecting the oxygen transfer from the surface, $K_L a$ can be shown to be:

$$K_L a = \frac{F'' H^{2/3} G_s}{V d_B} \tag{3-23}$$

Equation (3-23) is applicable to aeration depths in excess of 3 ft.

Equation (3-19) can be combined with Equation (3-23) for any operating temperature:

$$K_L a \cdot V = F''' G_s^{(1-n)} H^{(1-g)} \tag{3-24}$$

Equation (3-24) can be employed to characterize the performance of diffused aeration devices.

Example 3-3. In a particular aeration system $K_L a$ was measured as 14·0/hr at a gas flow of 80 cm/min. If the bubble diameter is 0·182 cm and the velocity 26·5 cm/sec, compute K_L when

$$\text{depth} = 300 \text{ cm}$$

$$\text{volume} = 4750 \text{ cm}^3$$

$$A/V = \frac{6 G_s H}{d_B V_B V}$$

$$A/V = \frac{6 \times 80 \times 300}{60 \times 0.182 \times 26.5 \times 4750} = 0.105/\text{cm}$$

$$K_L = K_L a \frac{V}{A} = 14.0 \times \frac{1}{0.105} = 134 \text{ cm/hr}$$

Effect of Temperature on Oxygen Transfer

The aeration coefficients are influenced by temperature due to the effect on the diffusivity and viscosity. The effect of temperature on the coefficients is usually expressed as follows:

$$K_T = K_{20} \cdot \theta^{T-20} \qquad (3\text{-}25)$$

in which,

$T = $ temperature, °C

$K_T = $ coefficient at temperature, T

$K_{20} = $ coefficient at 20°C

The temperature coefficient, θ has been reported to vary from 1·016 to 1·047. Data and formulations have been reported by Haslam *et al.* (1924), Streeter *et al.* (1936) and Wilke (1949). In bubble aeration, temperature also influences the bubble diameter and velocity. It appears that the presence of surface active agents and other substances may also affect the temperature coefficient. Studies on bubble aeration indicated a temperature coefficient of 1.02. Temperature correlations reported by various investigators are referenced.

Example 3-4. If the aeration system in Example 3-3 is operating at 20°C, compute the transfer coefficient at 30°C

$$k_L a_{(30°C)} = K_L a_{(20°C)} \cdot 1\cdot02^{(t-20)}$$

$$= 1\cdot40 \cdot 1\cdot02^{10}$$

$$= 1\cdot71/\text{hr}$$

Effect of Waste Constituents on Oxygen Transfer

The effect of various constituents on the oxygen transfer coefficient has been recognized. Various data are available indicating the reduction in reaeration which is generally associated with solutions other than pure water. This condition has also been observed in bubble aeration. In this case, however, the effect is two-fold in that various constituents not only influence the transfer coefficient, but also the size of the bubble and therefore the ratio of the interfacial area to the volume. In the case of natural streams the ratio of the surface area to the volume is a constant for a given flow condition. The effect of the majority of waste substances on the reaeration coefficient is reflected by their influence on the diffusivity. Waste constituents, such as surface-active substances, concentrate at the liquid–air interface and apparently create a barrier to the diffusion of oxygen. At the interface a thin layer exists through which the diffusion is molecular in nature and below which the eddy diffusion controls. Since substances such as surface-active agents concentrate at the interface, the molecular diffusion is hindered. The magnitude of this influence is a function of concentration of the waste substance and, at relatively low concentrations, the reported reductions in reaeration rate are significant.

Bubble size decreases with decreasing surface tension, thereby increasing the interfacial area for transfer per unit volume. It has been reported that very small quantities of added surface-active agent can change the effective bubble size as much as 100 times. Visual observations on the aeration of peptone and syndets showed a change in bubble size from coarse in water to fine in the presence of the added constituent. Reduction in bubble size decreases the rate of surface renewal due to a slower rate of bubble rise and a reduced shear at the bubble boundary.

Recent studies on bubble aeration (Eckenfelder and Barnhart, 1960) have shown that the liquid film coefficient K_L in bubble aeration decreases rapidly in the presence of small concentrations of surface active agents. At high concentrations, k_L tends to increase slightly. The overall coefficient $K_L a$ shows an initial decrease followed by a sharp increase at high concentrations, primarily due to the increased interfacial area-volume ratio, A/V. Data for bubble aeration in heptanoic acid solution is shown in Fig. 3-4. The nature of the aeration surface exerts a significant effect on the transfer rate in the presence of surface active agents. Holroyd (1952) has shown that in the presence of surface active agents the most severe reduction in K_L occurs at the bubble surface. A less severe reduction occurred at a stagnant film surface and the least depression at a heaving water surface. At the heaving water surface it is postulated

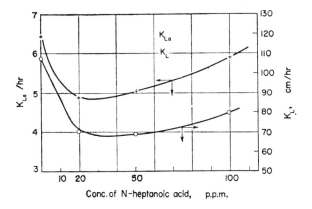

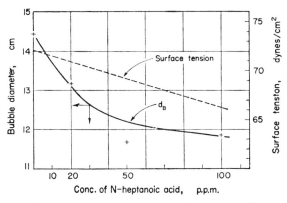

FIG. 3-4. Effect of various concentrations of heptanoic acid on oxygen transfer from air bubbles in water.

that the short life of any interface restricts the formation of an adsorbed film. Downing and Truesdale (1955) showed that the addition of 1 p.p.m. of alkylbenzene sulfonate to sea water reduced K_L from 10·25 to 5·1 cm/hr through a surface ruffled by air jets while the same concentration in tap water agitated by waves 8 cm high caused a reduction of only 21·6–16·3 cm/hr. Cullen and Davidson (1956) showed a decrease in K_L followed by an increase at high concentrations of surface active agents. The transfer rate, measured in a flow of water over a sphere, dropped to a minimum and then rose to approach the rate in pure water. They showed that the surface-active agents and impurities present constituted a resistance to mass transfer in the region where the surface tension was changing. This was attributed to the preferential adsorption of micelles at the interface at low concentration which were desorbed back into solution at high concentration. O'Connor (1960) has observed a similar phenomena employing an oscillating grid which simulates natural stream reaeration conditions and proposed an hypothesis relating this effect to the excess surface concentration, as defined by the Gibbs equation.

In order to compare the transfer rate in wastes to that in water, a coefficient α is defined as the ratio of $K_L a$, in waste to that in water under specified operating conditions. For bubble aeration from an aloxite diffuser α decreased from 1·0 in water to 0·5 in the presence of 25 p.p.m. of peptone. Peptone concentrations of 1000 p.p.m. reduced α to 20–40 per cent of that in water. Sawyer and Lynch (1954) found a variation in α of 0·35–1·0 with 50 p.p.m. of various anionic and non-ionic commercial detergents. Addition of a silicone antifoam to water reduced the absorption rate 35 per cent. Extensive studies by King (1955) showed α to vary from 0·26–0·46 for fresh sewage and from 0·16–0·19 for septic sewage. α was found to be 0·6, 0·7 and 1·4 for chipboard repulping wastes, kraft mill mixed wastes and semichemical paper machine wastes respectively. Bio-oxidation which alters the physical and chemical properties of waste mixtures modified α to a value approaching that of water. In activated sludge treatment of domestic sewage α increased from 0·72 to 0·90 after 4 hr of aeration. Treatment of kraft mill waste increased α from 0·45 to 0·79 after 3 hr of aeration with activated sludge. High mixed liquor solids concentrations in activated sludge processes reduce $K_L a$ by altering the bulk viscosity of the aerating medium. In the presence of 10,000 p.p.m. sludge solids, absorption was only 1/5 that in pure waste (Gaden, 1956). Kehr (1938) showed a reduction of K_2 in stream reaeration in the presence of oils, soaps and organic acids and raw and treated sewage.

MEASUREMENT OF OXYGEN TRANSFER COEFFICIENT

The performance of commercial aeration devices has been reported in the literature based on sulfite oxidation studies. When the tank geometry is similar and where data are available on the transfer characteristics of the waste relative to water, the sulfite data may be used for aeration design. In many cases, however, variations in the physical arrangement of the equipment and changes in the chemical nature of the waste necessitate evaluation of the oxygen transfer characteristics under actual operating conditions. The test procedures which may be employed for such a study are summarized below:

1. *Sulfite oxidation.* In the presence of copper or cobalt salts which act as catalysts, the reaction between sulfite ion and oxygen or air proceeds rapidly and irreversibly to completion in aqueous solution. The rate is independent of sulfite ion concentration between 0·015 and 1 M.

Water or waste is prepared 0·2 N in sulfite ion and 10^{-3} molar in cupric ion (Cobalt may be substituted for copper). Air is applied at predetermined rates and samples withdrawn at periodic time intervals (3–20 min) depending on the air flow and absorption rate. The rate of oxygen absorption is measured by determining the difference between the unoxidized sulfite ion concentration before and after aeration (Cooper *et al.*, 1944).

An aliquot sample depending on the unoxidized sulfite ion concentration is pipetted into each of two flasks containing 50 ml of 0·1 N iodine solution. During the transfer the tip of the pipette should be held close to the surface of the iodine solution to avoid aeration. The unoxidized sulfite ion is then determined by an iodiometric procedure of back titration with standard thiosulfate solution to a starch indicator endpoint. To minimize sulfite oxidation during transfer, the pipettes should be rinsed with distilled water and flushed with nitrogen for 2–3 min prior to sampling.

Results are expressed as p.p.m. O_2 absorbed per hour which is $K_L a \cdot C_s$.

Results obtained by the sulfite test must be interpreted with caution when considering design scaleup. Morgan (1958) showed that the oxygen transfer rate was a function of catalyst concentration below 5 p.p.m. of $CoCl_2$. Carpini and Roxburgh (1958) showed a higher temperature coefficient for sulfite oxidation than for oxygen transfer to deaerated water. Shultz and Gaden (1956) indicated that the controlling rate step may not be the mass transfer of oxygen to water but rather a step in the complex oxidation reaction for sulfite. In any case sulfite oxidation tests provide a convenient test for comparing aeration equipment under similar oxidation conditions.

2. *Non-steady state aeration.* In many cases it is desirable to determine the transfer coefficient $K_L a$ in aeration tanks without activated sludge. The dissolved oxygen present in the waste water can be removed using sodium sulfite to which cobalt has been added as a catalyst. Air is then admitted at the desired level and the dissolved oxygen measured at selected time intervals after the sulfite is oxidized. The transfer coefficient, $K_L a$, is computed from the slope of an experimental plot of $(C_s - C)$ vs. time of aeration. By repeating this procedure for various depths and air rates, the absorption characteristics for the system can be calculated.

3. *Steady-state aeration-activated sludge.* In the aeration of activated sludge, Equation (3-9) must be modified to account for the effect of oxygen utilization by the sludge–liquid mixture:

$$dc/dt = K_L a (C_s - C) - r_r \qquad (3\text{-}26)$$

in which the $r_r =$ oxygen uptake rate.

Under steady-state operation, $dc/dt = 0$ and Equation (3-26) becomes:

$$K_L a = \frac{r_r}{(C_s - C)} \qquad (3\text{-}27)$$

In those cases where the waste loading is reasonably constant and the oxygen uptake rate does not vary markedly with time, the transfer coefficient $K_L a$ can be evaluated from Equation (3-27). The dissolved oxygen in the tank liquor and the oxygen uptake rate can be measured by polarographic or chemical methods in the field. The saturation value employed in Equation (3-27) is computed for midpoint of the tank depth.

THEORY AND PRACTICE OF AERATION   97

The oxygen uptake rate is affected by the turbulence level maintained in the aeration tanks. Since the polarographic procedure does not reproduce tank turbulence conditions at high turbulence levels such as are encountered in turbine aeration systems, analysis of the exit gas by the oxygen gas analyzer or Orsat apparatus will more accurately record the uptake rate existent in the tank itself.

The average uptake rate is obtained by averaging the exit gas content at several points across the surface of the aeration tank.

4. *Non-steady state aeration-activated sludge.* An accurate measure of the transfer coefficient $K_L a$ can be obtained by a non-steady state procedure under stabilized operating conditions. When the air is turned off or reduced to a low level in the aeration tanks, the dissolved oxygen will approach zero through microbial respiration. When the dissolved oxygen is removed from the system the air rate is returned to its operating level and samples withdrawn at one-minute intervals for dissolved oxygen measurement. Sampling is continued until a steady state condition is approached. The oxygen utilization rate is determined as previously described. The $K_L a$ value for the system is computed by a graphical method employing Equation (3-26).

DIFFUSED AERATION

Schematic representation of aeration devices is shown in Fig. 3-5 and examples in Fig. 3-6.

There are three basic types of diffused aeration devices commercially available:

(A) Small orifice diffusion units such as porous media, plates or tubes constructed of silicon dioxide or aluminum oxide grains held in a porous mass with a ceramic binder. Competitive units include saran or nylon-wrapped tubes or bags. These units may be permanently placed in the bottom of an aeration tank or suspended from flexible joints along the sidewall of a tank. When air is diffused through these units, a helical or screw motion is imparted to the sludge–liquid mixture. High maintenance costs may be encountered in some waste applications due to orifice clogging. Standard porous diffuser units are designed to deliver 4–8 ft^3/min per unit. The absorption efficiency depends on the size of air bubbles released. This in turn depends on the type and porosity of the diffuser unit.

For Carborundum plates the diameter of air bubbles has been found to vary from 0·2 cm to 0·43 cm over a permeability range of 40–120 (King, 1955). Downing (1960) showed that the average bubble diameter varied from 0·17 to 0·25 over a permeability range of 10–30. In order to maintain adequate circulating velocities, a mini-

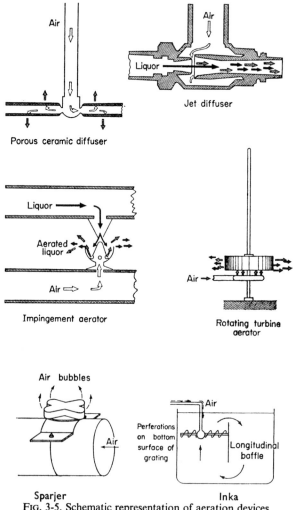

FIG. 3-5. Schematic representation of aeration devices (after Downing, 1960).

mum air flow of 3 cfm per lineal foot of tank must be maintained. The minimum spacing of units is 6 in. and the maximum spacing 2 ft. Diffuser plates are usually placed along one side of the bottom of an aeration tank and are designed to cover 5–10 per cent of the tank area.

Small orifice diffusion units are subject to clogging externally by materials in the tank liquor and internally by particulate matter in the air supply. The maximum recommended concentration of matter in the air supply is 0·1 mg/1000 ft^3. The rate of clogging and back pressure buildup increase with increasing relative humidity. External clogging results from ferrous iron oxidized to ferric iron, rust or scale from the air piping and deposition of organic solids or silt when the air supply is turned off. High concentrations of calcium carbonate or fine sand 100 μ or smaller will also increase the rate of clogging (Morgan, 1958).

(B) Units employing a mechanical or air shear such as the impingement or jet aerator. The impingement aerator employs a water stream air lifted from the aeration tanks as a shearing device for air bubbles discharged from a large orifice. The control variables are the impingement liquor flow, air flow and the location of the water nozzle relative to the air orifice. The necessity for air filters is usually eliminated.

The impingement type aerator is installed in headers up to 40 ft in length with impinger bowls and water nozzles saddle mounted to the air and water headers at from 15–24 in. centers. Each unit is designed to diffuse 4–16 ft^3/min of air with an impinger liquor flow of 15–20 gal/min. The bubble size released depends on the quantity of impingement liquor flow delivered by the circulating air lift pump. Absorption increases linearly with impingement liquor flow. Over the normal operating range, transfer efficiencies of 10–12 per cent can be expected based on sulfite oxidation tests.

The jet aerator pumps liquid from the aeration tanks through a piping manifold. The unit aspirates and disperses atmospheric air or air from a blower within the ejector's capacity. Power consumption is a consideration since the liquid is circulated at 25–30 lb/in^2 pressure. Orifice clogging problems may result in certain applications. A wide range of oxygen absorption can be obtained with the jet aerator depending upon the selection of operating variables. Studies by Kountz and Villforth (1954) showed that the rate of dissolved oxygen supplied is a function of the nozzle stream velocity as is also the air volume aspirated. For various units, 0·66–1·0 lb O₂/hr can be transferred at liquid pumping rates of 34–44 gal/min. These values can be increased by about 50 per cent by supplying additional air with a blower. 18·3–20·1 per cent absorption has been reported by Hauer (1955) in sulfite oxidation.

(C) Large orifice diffusion units such as the sparjer and discfuser. The sparjer contains four short tube orifices at 90° centers, from which air is emitted at high velocity. Tank turbulence tends to redivide large bubbles into small bubbles. The discfuser emits air from around the periphery of a disc. The transfer efficiency from these devices is related to tank geometry and the location of the diffuser unit. Efficiency may increase or decrease with gas rate depending on tank geometry.

Other recent types of air diffusion equipment include:

(a) The hydraulic shear diffusor which is a box into which air is discharged through an open pipe. The rising air–water mixture and the downward water flow create turbulence and hydraulic shear which breaks the large bubbles into smaller bubbles.

(b) The venturi diffusor in which an air–water mixture flows into a venturi section from which it discharges into the aeration tank. Most of the energy is expended in this device in accelerating water through the venturi section.

(c) The Inka system in which pipes in the form of a grating perforated on the underside are mounted about $2\frac{1}{2}$ ft below the liquid surface on one side of a longitudinal vertical baffle. Very high air flows under low pressure are used to maintain a spiral flow and to maintain the sludge in suspension.

Diffused Aeration Performance

The rate of oxygen transfer from diffused aeration devices is dependent on the nature of the diffusion device, the submergence depth and the gas flow rate. For comparative purposes it is convenient to express Equation (3-24) as an absorption number.

$$N = \frac{VK_L a}{G_s^{(1-n)} h^{(1-g)}} \tag{3-28}$$

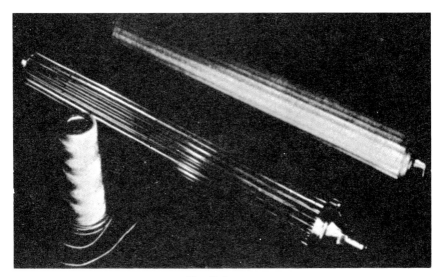

Fig. 3–6a.
Fig. 3-6 (a–d). Diffused aeration units, (*Courtesy of Chicago Pump Co.*)

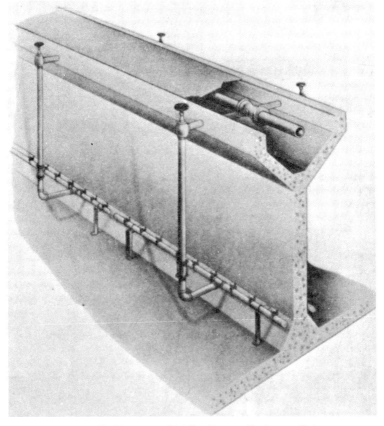

Fig. 3-6b. (*Courtesy of Walker Process Equipment Co.*)

Fig. 3–6d.
(Courtesy of Dorr-Oliver Inc.)

Fig. 3–6c.
(Courtesy of Walker Process Equipment Co.)

FIG. 3-10. Turbine aeration unit.
(*Courtesy of Pflaudler-Permutit Inc.*)

Transfer data for a sparjer unit in a circular aeration tank is correlated according to Equation (3-28) in Fig. 3-7. Values for typical diffusion units on water and various wastes are shown in Table 3-1. Performance of several diffusion devices in spiral flow tanks is shown in Fig. 3-8.

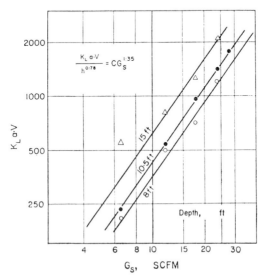

FIG. 3-7. Oxygen transfer characteristics of a Sparjer unit in a circular aeration tank.

Effect of Gas Rate on Oxygen Transfer

The volume of bubbles produced from small porous diffusion units as related to the orifice size. For small porous diffusion units (carborundum plates and tubes, seran tubes, etc.) the exponent $(1 - n)$ in Equation (3-28) is less than 1·0, indicating an increasing bubble size with increasing air rate. King (1955) showed that the bubble diameter changed 10 per cent over a range of air flow of 1 scfm to 4 scfm for carborundum plates. Over a wide range of bubble fineness and dispersion Hixon and Gaden (1950) showed the exponent $(1 - n)$ to vary from 0·33–0·82. For these devices, when $(1 - n)$ is less than 1·0, the absorption efficiency decreases with increasing air rate. The effect of air rate on oxygen absorption for several aeration devices is shown in Fig. 3-8.

In the case of the sparjer and discfuser units with increasing air rates, tank turbulence tends to redivide larger bubbles into smaller

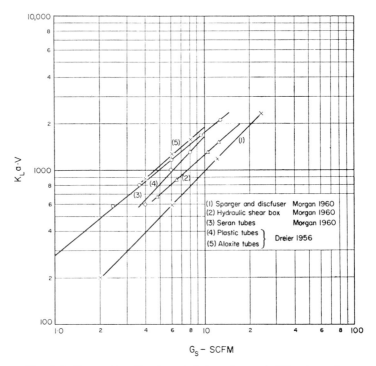

Fig. 3-8. Oxygen transfer characteristics of several diffused aeration devices.

bubbles increasing both the surface renewal rate and the A/V ratio. When this occurs, the gas rate exponent $(1 - n)$, may be greater than 1·0 resulting in an increasing transfer efficiency with increasing gas rate. This phenomenon, however, is influenced by the tank geometry and the location of the diffuser unit. It will be observed from Table 3-1 that sparjer units centrally located and in the sulfite test tank showed exponents greater than 1·0 while in the sewage and pulp and paper waste units, the exponent was less 1·0. The same

TABLE 3-1. OXYGEN TRANSFER CHARACTERISTICS OF DIFFUSED AERATION EQUIPMENT

Unit	System of Waste	Absorption number $k_L a V$ $G_s^{(1-n)}(1-n)$ h, ft			Description
Aloxite tubes*	Sulfite	270	0.85	15	3 units in test tank 12 ft × 5 ft; operating temp. 20° C; new clean tubes.
Plastic tubes*	Sulfite	160	1.0	15	
Sparjer*	Sulfite	61	1.35	15	
Impingement*	Sulfite	145	1.10	15	
Discfuser‡	Sulfite	92	1.20	13	Impingement liquor; flow 18 gal/min air flow range 8–16 scfm column 30 in. diam. × 13 ft depth‡.
Sparjer	Sewage	213	0.86	15	Aeration tank 30 ft width, baffled walls, high rate activated sludge; SS–2000 p.p.m.
Sparjer	Sewage	207	0.86	15	Two units placed on c of one-half of a 14 ft diameter tank.
Sparjer	Pulp and paper	57	1.35	15	
Discfuser	Pulp and paper	168	0.95	15	Aeration bay 116 ft length; surface area 2780 ft²; activated sludge 2000 p.p.m.; 100 units/bay.
Sparjer	Pulp and paper	279	0.81	15	
Impingement	Pulp and paper	195	1.10	15	
Impingement	Pulp and paper	108	1.10	15	Same as above but after six months service when units were partially clogged.
Carborundum plates†	Water	65	0.80	26	Tanks 137 ft × 22 ft × 26 ft deep; diffuser plate area of 12 per cent of tank area.
Jet unit	Pharmaceutical	710	1.00	4	Shutte-Koerting jet No. 00 SK; operation H₂O 3–5 gal/min at 20–65 ft head in 55-gal drum.
Sparjer§	Water	100	1.00	15	2 units in test tank 24 ft × 4 ft; operating temp. 20° C.
Seran tubes§	Water	275	0.8	15	
Hydraulic shear box§	Water	180	0.85	15	1 unit in test tank 24 ft × 4 ft; operating temp. 20° C.

* Dreier 1956.
† King 1955.
‡ Data of Chicago Pump Co.
§ Morgan (1959).

effects were reported for the discfuser by the Chicago Pump Co. This emphasizes the importance of relating absorption efficiency to actual tank conditions.

Morgan (1960) showed that the absorption characteristics of some diffusion devices may be influenced by the location of diffusors with respect to each other. For example, seran wrapped tubes at an air flow of 4 scfm per tube produced 14·3 per cent, 12·8 per cent and 12·2 per cent absorption efficiency at spacings of 2 ft, 1 ft and 0·67 ft respectively.

Effect of Depth on Oxygen Transfer

Over the range of depths employed in commercial aeration practice the overall coefficient $K_L a$ can be related to an exponential function of depth as shown in Equation (3-28). The exponent on depth $(1 - g)$, has been experimentally determined for several diffused aeration systems and is summarized in Table 3-2.

TABLE 3-2. EFFECT OF DEPTH ON THE OXYGEN TRANSFER RATE COEFFICIENT

System	Exponent (Eq. 3-28) $(1 - g)$
Diffusors in still water column	0·67
Plate diffusors*	0·71–0·77
Diffusair Sparjers	0·78
Aloxite Tubes†	0·45
Impingement†	0·65

* Date of King (1955).
† Date of Dreier (1956).

Effect of Waste Characteristics

The effect of waste characteristics on oxygen transfer is discussed previously. As bio-oxidation proceeds and the contaminants are destroyed by oxidation the transfer rate approaches that in water ($\alpha \to 1·0$). Typical data obtained from laboratory studies for various industrial wastes are shown in Table 3-3.

The oxygen transfer rate in industrial waste treatment systems has

THEORY AND PRACTICE OF AERATION 105

been observed to vary from day to day. In those cases where more than one set of results were available, this variation is indicated in Table 3-3. This variation can be attributed to changes in the chemical composition of the waste, variation in aeration solids and variation in the physical characteristics of the aeration system (temperature, pressure, etc.). The variation that was encountered in the aeration of pulp and paper mill wastes with activated sludge using sparjers, measured daily, over a period of 45 days showed that 10 per cent of the time $K_L a$ was equal to or less than 3·4 and 90 per cent of the time $K_L a$ was equal to or less than 8·0. This variation can be compensated for in practice by maintaining 2 p.p.m. dissolved oxygen or more under average conditions and a constant air flow. For the example previously cited at a 10 per cent frequency level the dissolved oxygen would be 0·5 p.p.m.

TABLE 3-3. SUMMARY OF OXYGEN TRANSFER CHARACTERISTICS OF SOME INDUSTRIAL WASTES AND THEIR BIO-OXIDATION EFFLUENTS

Waste	BOD		α	
	Raw	Effluent	Raw	Effluent
Paper repulping*	187	50	0·68	0·77
Semi-chemical* machine back-water	1872	—	1·40	—
Mixed kraft mill	150–300	37–48	0·48–0·86	0·70–1·11
Pulp and paper (bleach plant)†	250‡	30‡	0·83–1·98	0·86–1·0
Pulp and paper (pulp mill)†	205‡	—	0·66–1·29	—
Pharmaceutical	4500	380	1·65–2·15	0·73–0·83
Domestic sewage (fresh)	180	9	0·82	0·98
Synthetic fibre	5400	585	1·88–3·23	1·04–2·65
Board mill	660‡	—	0·53–0·64	—

* Paper repulping and semi-chemical wastes mixed prior to bio-oxidation.
† Bleach plant and pulp mill wastes mixed prior to bio-oxidation.
‡ Average values.

Tank Circulating Velocities

The rate of interfacial renewal in an aeration tank is proportional to the velocity of the bubble relative to the tank liquid. High circulating velocities will produce large values of the liquid film coefficient, K_L. In addition, interfacial renewal rates at the tank surface will be related to velocity gradients at the surface.

The horizontal velocities in aeration tanks for several commercial diffusion devices are summarized in Table 3-4. The surface velocity

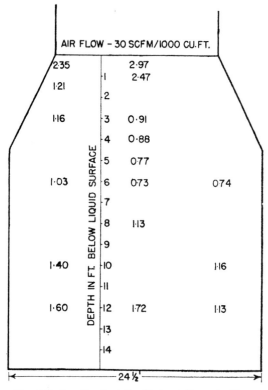

FIG. 3-9. Aeration tank liquid circulating velocities.

measured by King (1955) for the condition shown in Table 3-4 in a 34 ft wide aeration tank was found to vary from 4·3 ft/sec above the diffuser plates to 2·1 ft/sec near the side wall opposite the plates. A velocity profile obtained in an activated sludge aeration tank treating pulp and paper mill wastes is shown in Fig. 3-9. The data shown in Fig. 3-9 were obtained by lowering a current meter to the indicated depths. The liquid velocities near the surface ranged from 2·0 to 4·0 ft/sec and from 0·5 to 2·0 ft/sec at greater depths.

TABLE 3-4. VELOCITY VARIATIONS WITH AERATION DEVICES

Depth ft	Pulp and paper waste Sparjers 30 scfm/1000 ft^3 ft from center line			Pulp and paper waste Impingement 28 scfm/1000 ft^3 ft from center line		
½	2·35	2·97	—	—	2·23	—
1	—	2·47	—	—	1·40	1·36
1½	1·21	—	—	—	—	1·10
2	—	—	—	—	0·97	1·05
3	1·16	0·91	—	0·74	0·88	0·74
4	—	0·88	—	0·75	0·63	1·10
5	—	0·77	—	0·70	0·50	0·79
6	1·03	0·73	0·74	0·47	0·52	0·67
8	—	1·13	—	0·49	0·70	0·85
10	1·40	—	1·16	0·51	0·71	1·13
12	1·60	1·72	1·13	—	1·10	1·05
14	—	—	—	—	1·44	1·23

* Aeration tank 15 ft depth; 2·45 ft width.
† Aeration tank 15 ft depth; 34·75 ft width; data of King (1955).

A traverse made of dissolved oxygen values at various points in the aeration tank described above showed the dissolved oxygen values substantially constant throughout the tank. It is therefore reasonable to expect that at the turbulence levels encountered in commercial aeration practice, no concentration gradient will exist in the aeration tank itself.

Oxygen Transfer Efficiency

It is practically significant when comparing aeration devices or when evaluating absorption in various wastes to consider the oxygen transfer efficiency.

The transfer efficiency is computed in the following fashion:

$$\% \text{ eff.} = \frac{\text{wt. O}_2 \text{ absorbed/unit time}}{\text{wt. O}_2 \text{ supplied/unit time}} \times 100$$

in which:

$$R_d = K_L a (C_s - C) \cdot 8{\cdot}34 \cdot \text{mg of tank capacity}$$

$$= r_r \cdot 8{\cdot}34 \cdot \text{mg of tank capacity}$$

$$= \text{lb } O_2/\text{hr}$$

and

$$G_s = \text{air flow (cfh)} \cdot \text{density } \frac{\text{lb air}}{\text{ft}^3} \cdot 0{\cdot}232 \frac{\text{lb } O_2}{\text{lb air}}$$

the air density is related to the temperature and pressure of the air

$$\rho = 0{\cdot}0808 \left(\frac{P_b}{14{\cdot}7}\right)\left(\frac{492}{T_f}\right)$$

In which T_f is absolute temperature.

The air temperature will rise due to adiabatic compression of dry air. An increase to 8 p.s.i./gauge pressure will increase the air temperature from 68°–73°F depending on the atmospheric temperature.

Example 3-5. Compute the oxygen absorption efficiency at 1·0 p.p.m. dissolved oxygen in a waste oxidation system. C_s is 7·77 p.p.m. and the air flow is 3400 cfm. The tank capacity is 1·41 mg. $K_L a$ is 4·89.

$$\rho = 0{\cdot}0808 \left(\frac{22{\cdot}7}{14{\cdot}7}\right) \cdot \left(\frac{492}{610}\right) = 0{\cdot}1$$

$$\% \text{ eff.} = \frac{4{\cdot}89 \cdot 6{\cdot}77 \cdot 8{\cdot}34 \cdot 1{\cdot}41}{3400 \cdot 60 \cdot 0{\cdot}232 \cdot 0{\cdot}1} \cdot 100 = 8{\cdot}25\%$$

The absorption efficiency for the units shown in Table 3-1 at near optimum air rates are shown in Table 3-5.

The absorption efficiency will vary depending on the nature of the waste and the geometry of the aeration tank. For example, at 16 scfm/unit, sparjers produced 10 per cent absorption in sulfite in a rectangular tank. In sewage oxidation in a spiral flow tank the absorption was 8·6 per cent at the same air flow. The transfer ratio α

TABLE 3-5. OXYGEN ABSORPTION EFFICIENCY OF AERATION DEVICES

Unit	Waste	Depth, ft	Air rate scfm/unit	% absorption (zero dissolved oxygen)
Aloxite tubes	Sulfite	15	6	13·7
Plastic tubes	Sulfite	15	6	10·0
Sparjer	Sulfite	15	16	10·0
Impingement	Sulfite	15	12	11·7
Discfuser	Sulfite	13	16	9·8
Sparjer	Sewage	15	16	8·6
Sparjer	Pulp and paper	15	16	9·0
Sparjer*	Pulp and paper	15	16	9·8
Discfuser	Pulp and paper	15	16	8·6
Jet	Pharmaceutical	4	0·5	24·8
Sparjer	Water	15	16	6·4
Seran tube	Water	15	6	12·5

* Activated sludge plant.

for this sewage could be expected to range from 0·75–0·85. In a pulp and paper waste ($a = 0·9$–$1·0$) in a spiral flow aeration tank the absorption was 9·8 per cent. Similar observations can be made of the other types of aeration devices.

Diffused Aeration Design

In diffused aeration systems $K_L a$ may be varied by:

(a) Unit spacing selection. Minimum spacing is limited by interfering air diffusion patterns and maximum spacing by the necessity to maintain adequate tank circulation velocities.

(b) Bubble size. Bubble size in porous diffuser units depends on permeability while in the impingement type unit on the circulating liquor flow.

(c) Air flow.

A suggested design procedure for aeration systems in biological oxidation processes is outlined below:

1. Compute the oxygen saturation characteristics, C_s, at process operating temperatures and pressure for the particular waste to be treated according to Equation (3-4).

2. The minimum operating dissolved oxygen level should be maintained between 0·5 and 1·0 p.p.m. to insure aerobic action.

3. The BOD removal requirements are usually established by stream conditions or state standards. Aeration requirements and biological solids level in the aeration tanks are defined by laboratory or pilot plant studies. The oxygen uptake rates for various BOD removal levels can be estimated from formulations previously derived.

4. The oxygen demand distribution will be a function of time of aeration. This may be obtained by a laboratory study. The actual demand distribution in the aeration tanks will depend on the hydraulic characteristics of the particular tanks as designed.

5. $K_L a$ for each section of the system can be computed from Equation (3-27).

6. The operating $K_L a$ derived from Equation (3-27) must be corrected for temperature if the operating temperature deviates from 20°C and for the oxygen transfer coefficient, a, of the specific waste.

7. From Fig. 3-7 (or a similar plot) the unit air flow–volume factor to transfer the required oxygen is selected or computed from Equation (3-27).

8. After selecting a tank width and an air flow/unit, the unit spacing is computed.

9. This should establish the design for the most severe operating conditions. Under less severe conditions (winter operation, lower BOD loadings, etc.), the reduced required air flow can be computed in a similar manner.

Example 3-6. (a) *Oxygen uptake rate calculation*

The average oxygen uptake rate in an aeration tank is 36·8 p.p.m./hr. The average uptake in each quarter section of the tank is:

Tank length %	Uptake rate	
	Ratio to mean	p.p.m./hr
0–25	1·3	1·3 . 36·8 = 47·9
25–50	1·0	1·0 . 36·8 = 36·8
50–75	0·8	0·8 . 36·8 = 29·5
75–100	0·7	0·7 . 36·8 = 25·8

THEORY AND PRACTICE OF AERATION

(b) *Calculation of average α*

$$\alpha \text{ for raw waste} = 0{\cdot}60$$

$$\alpha \text{ for effluent} = 0{\cdot}90$$

If the recycle is 30 per cent the average α at the influent end of the tank is

$$0{\cdot}7 \cdot 0{\cdot}6 + 0{\cdot}3 \cdot 0{\cdot}9 = 0{\cdot}69$$

If the increase in α through the aeration tank is linear, the mean in the first quarter of the tank is 0·72.

(c) *Aeration tank dimensions*

The total tank volume is 0·4 mg (53,300 ft³). Assuming a depth of 15 ft and width of 20 ft, the length is 175 ft (44 ft/quarter).

(d) *Oxygen saturation*

Operating temperature = 30° C; assuming transfer efficiency 8 per cent; submergence on diffusers = 13 ft.

$$C_s = fC_w \left(\frac{Pb}{29{\cdot}4} + \frac{Ot}{42}\right) = 7{\cdot}25 \left(\frac{20{\cdot}3}{29{\cdot}4} + \frac{19{\cdot}6}{42}\right) = 8{\cdot}4 \text{ p.p.m.}$$

(e) *Required $K_L a$—first quarter*

Employing an operating dissolved oxygen level of 1·0 p.p.m.

$$K_L a\ 30° = \frac{r_r}{C_s - C_L} = \frac{47{\cdot}9}{(8{\cdot}4 - 1{\cdot}0)} = 6{\cdot}47$$

$$K_L a\ \text{at}\ 20° = \frac{K_L a\ (30°)}{1{\cdot}016^{T-20}} = \frac{6{\cdot}47}{1{\cdot}17} = 5{\cdot}53 \text{ per hr}$$

Correcting for α

$$K_L a = \frac{5{\cdot}53}{0{\cdot}72} = 7{\cdot}69 \text{ per hr}$$

The performance of a particular air diffuser in water at 20°C can be expressed by the relationship in Equation (3-28)

$$31 \cdot 3 = \frac{K_L a \cdot V}{G_s^{0.9} H^{0.7}}$$

For an air flow of 8 scfm and a submergence depth of 15 ft:

$$K_L a \, V = 1350$$

$$V = \frac{1350}{7 \cdot 69} = 176 \text{ ft}^3$$

The unit spacing for the diffusers is:

$$\frac{176}{15 \times 20} = 0 \cdot 59 \text{ ft}$$

The second, third and fourth quarter may be computed in a similar fashion. If the computed spacing is beyond design limits the unit depth and width must be adjusted to bring the unit spacing within the specification.

MECHANICAL AERATION

Mechanical aerators entrain atmospheric oxygen by surface agitation or disperse compressed air by a shearing and pumping action employing a rotating turbine or agitator. In the latter unit, air bubbles are discharged from a pipe or sparge ring beneath the agitator and are broken up by the shearing action of the high-speed rotating blades of the agitator moving through the liquid. The relative effect of liquid pumping and air shear depends upon the size and type of the agitator with respect to the tank. For systems of low oxygen utilization rate, oxygen may be supplied by air self-induced from a negative head produced by the rotor. This eliminates the necessity for external blowers or compressors.

The performance of an impeller system using compressed air can be defined by the generalized relationship:

$$K_L a V = c N^x G_s{}^y d_t{}^z \tag{3-29}$$

in which

c = constant characteristic of aerator
N = impeller peripheral speed
d_t = impeller diameter

Depending on the impeller geometry the exponents x, y and z have been found to vary from 1·2 to 2·4, 0·4 to 0·9 and 0·6 to 1·8 respectively.

The horsepower requirements of a turbine aerator can be expressed by the relationship:

$$HP = c d_t^n N^m \tag{3·30}$$

The exponents n and m generally vary from 4·8 to 5·3 and 2·0 to 2·5 respectively. The actual drawn horsepower decreases as air is increased under the impeller due to the decreased density of the aerating mixture. This relationship is shown in Fig. 3-11.

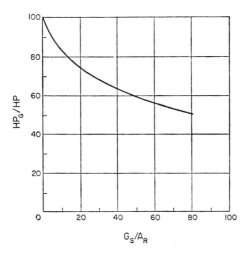

FIG. 3-11. Effect of air rate on turbine horsepower.

The performance of a dual impeller system treating pulp and paper mill wastes in which the top impeller was located 13 in. below the liquid surface is summarized in Table 3-6 (Laws and Burns, 1959).

TABLE 3-6*

Bay No.	O_2 utiliz. rate p.p.m./hr	Lb O_2/hr absorbed	Horsepower applied/bay			Lb O_2 absorbed /h.p. hr	Air flow scfm	% Eff. of turbine aerators
			Compressor	Turbine	Total			
1	98	309	62·3	49·3	111·6	2·77	1470	20·2
2	35	124	37·0	18·8	55·8	2·23	880	13·8
3	23	87	19·4	16·8	36·2	2·40	460	18·3
4	17	66	9·8	18·0	27·8	2·37	230	28·3

* All data compiled from 17 determinations on one of two units.

Example 3-7. A turbine aerator in an aeration tank 30 ft × 50 ft × 15 ft was found to transfer oxygen according to the relationship:

$$K_L a \cdot V = 25 \, G^{0·45} N^{1·5} d^{1·8}$$

The saturation of oxygen in the sewage liquid at 20°C is 8·45 p.p.m. The average saturation at the midpoint of the tank depth is 9·3 p.p.m. (Equation 3-4).

(a) Compute the oxygen transfer from a 40 in. diameter turbine rotating at 15 ft/sec peripheral speed with an air flow of 300 scfm.

$$K_L a \, V = 25 \, (300)^{0·45} \, (15)^{1·5} \, (3·33)^{1·8}$$

$$= 164{,}000 \text{ ft}^3/\text{hr}$$

$$V = 15 \times 30 \times 50 = 22{,}500 \text{ ft}^3$$

$$K_L a = 164{,}000/22{,}500 = 7·3/\text{hr}$$

THEORY AND PRACTICE OF AERATION

The total oxygen transferred per hour is:

$$\text{lb } O_2/\text{hr} = K_L a \cdot C_s \cdot V \cdot 8.34$$

$$= (7.3)(9.3)(0.169)(8.34)$$

$$= 96 \text{ lb/hr}$$

(b) Compute the required turbine horsepower

The ungassed horsepower for the above turbine is defined by the relationship:

$$\text{h.p.} = 0.02 \, d^{5.25} \, n^{2.75}$$

in which n is expressed in rev/sec

$$\text{h.p.} = 0.02 \, (3.33)^{5.25} \, (1.43)^{2.75}$$

$$= 29.5$$

(c) Compute the lb O_2 transferred/h.p. hr.

When air is discharged beneath the rotating blades of a turbine, the decreased density of the air–liquid mixture reduces the drawn horsepower. This relationship is shown in Fig. 3-11.

For the turbine in Example 3-7 the impeller area is:

$$A = \frac{\pi}{4} (3.33)^2 = 8.7 \text{ ft}^2$$

$$G_s/A = \frac{300}{8.7} = 34.5$$

From Fig. 3-11

$$\frac{\text{h.p.}_G}{\text{h.p.}} = 0.65$$

and the drawn turbine horsepower is:

$$29.5 \times 0.65 = 19.2 \text{ h.p.}$$

The blower horsepower can be computed

$$\text{h.p.} = \frac{G_s(\text{scfm})\, p(\text{lb/in}^2)\, 144}{33{,}000 \times E}$$

and if

$$E = 0\cdot 65;\ p = 5\cdot 55\ \text{lb/m}^2;\ \text{h.p.} = 11\cdot 2$$

The transfer efficiency is:

$$\frac{96\ \text{lb}\ O_2/\text{hr}}{30\cdot 4\ \text{h.p.}} = 3\cdot 16\ \text{lb}\ O_2/\text{h.p.-hr}$$

Cavitator

In the cavitator unit a submerged rotor is moved through the liquid at high velocities. Pressures at the rotor surface are reduced below atmospheric and a cavitation zone occurs, into which air is drawn through nozzles at the ends of the rotor. Fine bubbles are formed which are circulated through the tank liquid. Results from tests conducted in a tank $8 \times 8 \times 9$ ft deep employing the sulfite method at 7 ft 2 in. rotor submergence are summarized below (Hurwitz, 1959):

scfm Air	lb/hr O_2 Absorbed	% O_2 Absorption	KWH lb O_2 abs.
5·0	2·64	47·4	0·30
7·5	3·57	42·6	0·32
15·0	5·88	36·2	0·57

Simplex Hi-Cone

The Simplex Hi-Cone unit consists of a revolving cone through which large volumes of liquid are lifted through the updraft tube and discharged over the lip of the cone in a spray over the surface of the tank liquid. The liquid spirals from the cone toward the outer edges of the tank. The liquid tank contents are circulated through the draft tube of the cone (Fig. 3-12).

THEORY AND PRACTICE OF AERATION 117

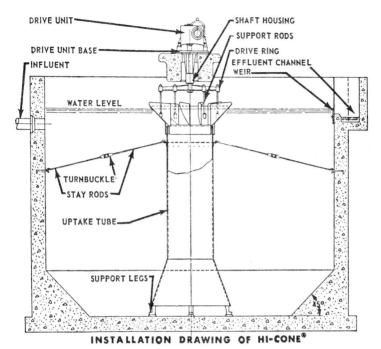

FIG. 3-12. Simplex Hi-Cone Unit (*Courtesy of Yeomans Brothers Inc.*).

TABLE 3-7. PERFORMANCE DATA OF THE HIGH CONE UNIT

Cone speed rev/min	Freeboard in.	Oxygen transferred lb/hr	KWH/lb O_2 Absorbed	
35·0	6	9·86	0·258	Non-steady state using final
41·3	6	24·8	0·210	sewage effluent $t = 15°$ C
45·0	6	32·8	0·200	at zero D.O. Cones 6 ft
49·5	6	35·3	0·266	diameter. Manchester Report (1959).
26·4	4·5	6·2	0·310	Modified High Intensity
32·0	4·5	10·8	0·266	Cones. 6 ft diameter
34·1	4·5	13·0	0·280	Downing, 1959.
37·3	4·5	14·5	0·280	
30	4	10·5	0·30	Hurwitz, 1960. $T = 20°$C.
30	2	11·0	0·37	
30	0·5	11·9	0·35	

Increasing the speed of rotation of the cone from 26–37 rev/min in a 6 ft cone gave a three-fold increase in oxygenation capacity. The effect of free board on oxygenation capacity is not a primary variable, being less than 30 per cent under optimum conditions

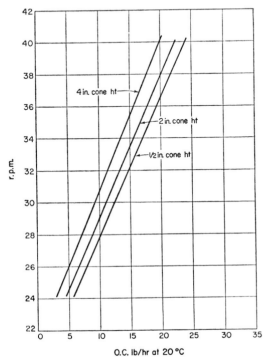

FIG. 3-13. Oxygen transfer characteristics of the Simplex Hi-Cone unit.

Downing, 1959). Some reported performance data are shown in Table 3-7 and in Fig. 3-13.

The Vortair surface aerator, as developed by Infilco, involves the use of a specially designed turbine located a short distance below the liquid surface. The oxygen entrainment from atmospheric air takes place due to two basic hydraulic phenomena that occur when the turbine is rotated at a proper speed. First, the radial discharge from the turbine, issuing slightly below the normal liquid level, produces a peripheral hydraulic jump, which has all the basic characteristics of a hydraulic jump including intense turbulence and air entrainment.

In addition to the air that is entrained in the peripheral hydraulic jump, a hydrodynamic condition is set up inside the turbine so that air is drawn into the flow issuing from the turbine. Under proper submergence conditions, for any given turbine diameter and speed, the top plate of the turbine becomes entirely free of liquid. Under

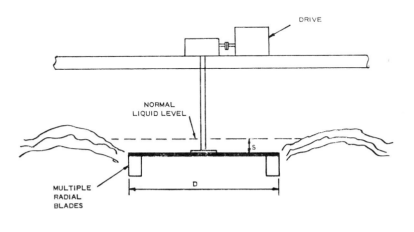

Fig. 3-14. Surface aeration unit (*Courtesy of Infilco Inc.*).

these conditions, air is sucked down past the top plate of the turbine due to the region of low pressure behind each of the radial vanes. The high-velocity water issuing radially from the turbine entrains air from the masses of air behind each of the radial blades, and thus the flow issuing from the turbine is full of fine bubbles.

Measurements made under controlled conditions have indicated that it is possible to dissolve over 6 lb of oxygen per hour per input horsepower to the turbine, under conditions when the liquid has zero dissolved oxygen, and when the liquid temperature is of the order of 85°F or less. The inherent liquid pumping characteristics of the turbine permit the movement of large quantities of liquid from the bottom of the basin into the oxygenating region surrounding the turbine. These aerators are capable of mixing the contents of large basins at depths varying from 15 ft to a few feet. Heavy concentrations of activated sludge can be kept in suspension.

REFERENCES

1. ADENEY, W. E. and BECKER, H. G., *Phil. Mag.*, 38 (1919) and 39 (1920).
2. CARPINI, R. E. and ROXBURGH, J. M., *Canad. J. Chem. Engng.* 36, 2, 73 (April 1958).
3. City of Manchester, England, Rivers Dept. Report (1959).
4. CAPPOCK, P. D. and MICKLEJOHN, G. T. *Trans. Inst. Chem. Engrs.*, 29, 75 (1951).
5. COOPER, C. M., FERNSTROM, G. A. and MILLER, S. A., *Ind. Eng. Chem.*, 36, 504 (1944).
6. CULLEN, E. J. and DAVIDSON, J. F., *Chem. Eng. Sci.*, 6, 2 (1956).
7. DANCKWERTZ, P. V., *Ind. Eng. Chem.*, 32, 6, 1460 (1951).
8. DOWNING, A. L. and TRUESDALE, G. A., *J. Appl. Chem.*, 5, 570–581 (1955).
9. DOWNING, A. L., *J. Inst. of Publ. Hlth. Engrs.*, (London) (April 1960).
10. DREIER, D. E. *Biological Treatment of Sewage and Industrial Wastes* Vol. I, (Ed. by MCCABE, B. J. and ECKENFELDER, W. W.) Reinhold Pub. Corp., New York, N.Y. (1957).
11. ECKENFELDER, W. W., *Sew. and Ind. Wastes*, 31, 60 (1959).
12. ECKENFELDER, W. W., *Proc. Amer. Soc. Civil Engrs.*, *S.A.*, 4, 2090 (1959).
13. ECKENFELDER, W. W. and BARNHART, E. L., paper presented Amer. Inst. of Chem. Engrs. Atlanta, Ga. (Feb. 1960).
14. GADEN, E., *Biological Treatment of Sewage and Industrial Wastes* Vol. 1, (Ed. by MCCABE, B. J. and ECKENFELDER, W. W.) Reinhold Publ Corp., New York, N.Y. (1957).
15. GAMESON, A. H. and ROBERTSON, H. B., *J. Appl. Chem.*, 5, 503 (1955).
16. HASLAM, R. T., HERSHEY, R. L. and KEEN, R. H., *Ind. Eng. Chem.*, 16, 1225 (1924).
17. HAUER, R., Aeration Symposium, *Proc. 10th Ind. Waste Conf.*, Purdue Univ., (1955).
18. HIGBIE, J., *Trans. Amer. Inst. Chem. Engrs.*, 31, 365 (1935).
19. HIXON, W. and GADEN, E., *Ind. Eng. Chem.*, 42, 1768 (1950).
20. HOBERMAN, W. L. and MORTON, R. K., *Trans. Amer. Soc. of Civil Engrs.*, 121, 227 (1956).
21. HOLROYD, A., *Water and San. (Brit.)* 3, 301 (1952).
22. HURWITZ, E., *Oxygen Absorption Studies of the Cavitator System* Yeomans Bros. Co., Eng. Bulletin. (1959).
23. IPPEN, A. T., CAMPBELL, L. G. and CARVER, C. E., M.I.T. Hydrodynamics Laboratory, Tech. Rept. No. 7 (May 1952).
24. IPPEN, H. T. and CARVER, C. E., M.I.T. Hydrodynamics Laboratory Tech. Rept. No. 14 (1955).
25. KEHR, R. W., *Sew. Wks. Jour.*, 10, 2, 228 (1938).
26. KING, H. R., *Sew. & Ind. Wastes*, 27, 10, 1123 (1955) ; 27, 8 (1955) ; 27, 9 (1955).

27. KOUNTZ, R. R. and VILLFORTH, J. C., *Proc. 9th Ind. Waste Conf.*, Purdue Univ. (1954).
28. LAWS, R. and BURNS, O. B., Tech. Rept., West Virginia Pulp and Paper Co., Covington, Va. (1959).
29. LEWIS, W. K. and WHITMAN, W. C., *Ind. Eng. Chem.*, **16**, 1215 (1924).
30. MORGAN, P. F., *Proc. Paper* 1069 *S.E.D.*, *Amer. Soc. Civil Engrs.*, **84**, No. SA 2 (April 1958).
31. MORGAN, P. F. and BEWTRA, J. K. *Proc. 3rd Biological Waste Treatment Conference*, Manhattan College, N.Y. (1960).
32. O'CONNOR, D. J. and DOBBINS, W. E., *Trans. ASCE*, **123**, p. 641 (1958).
33. O'CONNOR, D. J., *Proc. 3rd Biological Waste Treatment Conf.*, Manhattan College, N.Y. (1960).
34. OLDSHUE, J., *Biological Treatment of Sewage and Industrial Wastes*, Vol. I, Reinhold Pub. Corp., New York (1956).
35. SAWYER, C. N. and LYNCH, W. O., *Sew. & Ind. Wastes*, **26**, 10, 1193 (1954).
36. SCOULLER, W. D. and WATSON, W., *Surveyor*, **86**, 2215 (1934).
37. SHULTZ, J. S. and GADEN, E. L., *Ind. & Eng. Chem.*, **48**, 12, 2209 (1956).
38. STREETER, H. W., WRIGHT, C. T. and KEHR, R. W., *Sew. Wks. Jour.*, **8**, 2, 282 (1936).
39. WILKE, C. R., *Chem. Eng. Prog.*, **45**, 218 (1949).

CHAPTER 4

STREAM AND ESTUARY ANALYSIS

THE concentration of dissolved oxygen is one of the most significant criteria in stream sanitation. The discharge of organic impurities, such as municipal sewage and industrial wastes, into a body of water presents a problem of primary importance in this regard. The decomposition of the organic matter by bacteria results in utilization of dissolved oxygen. The replacement of oxygen by reaeration occurs through the water surfaces exposed to the atmosphere. An increase in the concentration of organic matter stimulates the growth of bacteria and the oxidation proceeds at an accelerated rate. The concentration can be so great that there results a condition in which the receiving water body is completely devoid of dissolved oxygen. Every stream then is limited in its capacity to assimilate organic wastes; but, within this limit, there is a no more economical method of waste disposal. Evaluation of the natural purification capacity of a stream, therefore, is of fundamental and practical value. The stream may be regarded as a natural treatment plant. It is only necessary to determine the quantity of organic wastes which can be discharged to this plant, without impairing its efficiency and without destroying its usage for other purposes. The natural purification capacity of the receiving stream is the logical basis for the determination of the degree of waste treatment.

The concentration of many physical characteristics and chemical substances may be calculated directly, knowing the relative volumes of the waste stream and river flow. Chlorides and mineral solids fall into this category. Some substances in waste discharges are chemically or biologically unstable and their rates of decrease can be predicted or measured directly. Sulfites, nitrites, some phenolic compounds and BOD are examples of this type of waste. These simple relationships, however, do not apply to the concentration of dissolved oxygen. This factor depends not only on the relative dilutions, but also upon the rate of oxidation of the organic material and the rate of reaeration of the stream. As a parameter of pollution, this factor means very little in itself, only in its relation to reaeration characteristics and the dissolved oxygen levels is it significant.

In order to proceed with a complete analysis, it is necessary to collect the following field data at selected river stations:

(1) Biochemical oxygen demands, oxygen uptake rates, dissolved oxygens and temperatures.
(2) River and waste discharges, cross-sections and velocities during the period of the sampling.
(3) Supplementary data on suspended solids, volatile matter, sludge deposits, pH and chemical constituents may be required, depending upon the nature of the wastes and the condition of the river.

Analysis of these data is then required to determine the rate of deoxygenation (oxygen demands) and the rate of reaeration (oxygen resource). The interrelationship between these rates provides the basis for the determination of the pollution capacity of the river.

OXYGEN DEMANDS

The oxygen demand of a waste is quantitatively evaluated by the biochemical oxygen demand test, as described in Chapter 1.

The organic matter in a flowing stream is usually not oxidized in the same fashion as that in the laboratory BOD test. The difference between the two rates can be generally attributed to the difference in the biochemical and physical characteristics of each environment. In order to evaluate the reaction constant for the stream it is necessary to determine the BOD of river samples at successive stations downstream from the source of pollution. The times of passage from the waste discharge to these stations are also required. A plot of these data establishes a decreasing pattern of organic matter remaining in time, in accordance with Equation (1-4). The determination of the stream rate may be obtained by re-expressing this equation as follows:

$$K_r = \frac{1}{t} \log \frac{L_A}{L_B} \qquad (4\text{-}1)$$

Since the river value, K_r, is a measure of the rate of removal of BOD and not necessarily oxidation, L_A and L_B refer respectively to the BOD at upstream and downstream stations. Any such river BOD is itself a point on a laboratory BOD curve. It is necessary to define this curve for each station or at least for a few stations in order to compute the ultimate value from the 5-day value. If each BOD curve from the individual stations is defined by approximately the same reaction constant, then it follows mathematically that the rate is also first-order in the river but not necessarily of the same magnitude. If, on the other hand, the reaction constant varies for each station then the river rate will probably be a second or some intermediate order. Usually the first-order reaction is sufficiently representative of most river BOD data. From this discussion, it follows also that the seed material is of minor consequence—i.e. sewage or river water. Even if the sewage seed resulted in a higher 5 day BOD value (a higher K_1), the ultimate should be approximately the same, and it is this value which is of significance in stream analysis.

The stream rate is determined from an analysis of the BOD data taken at a series of river stations downstream from the source of pollution. The curve describing the decrease in BOD in the river is defined by a series of BOD values, each one of which is a point on a laboratory BOD curve itself. This condition is shown diagrammatically in Fig. 4-1. Most stream rates are higher than those determined in the laboratory. Furthermore it may be seen that the standard 5-day values are not necessary to define the stream rate. Any convenient time interval may be used provided it is consistent for all sampling. However, when the longitudinal mixing is pronounced, the apparent river rate as determined by Equation (4-1) may be less than the actual oxidation and laboratory rate. This effect is due to the short circuiting of portions of the waste which move downstream in times less than that of the theoretical detention time. Over short stretches, if the longitudinal mixing is sufficiently intense, a uniform concentration of BOD may result. This condition is characteristic of the uniformity of BOD in the aeration tanks of activated sludge plants. In this case, Equation (4-1) would not be representative of the oxidation process.

In considering the difference between the laboratory and stream rates, distinction should be made between those factors which affect the oxidation of organic matter and those which affect the removal of organic matter (Thomas, 1948). The former have a direct influence on the dissolved oxygen concentration while the latter may or may not have an effect. The characteristics of the

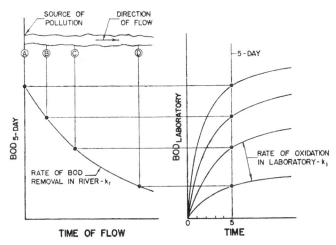

FIG. 4-1. River and laboratory BOD curves.

stream and the waste which influence the oxidation rate are as follows:

1. *Turbulence*—Turbulence increases the speed of many chemical reactions and it is probable that it has the same effect on biochemical reactions by increasing the opportunity of contact between organic matter and bacteria.
2. *Biological growths on stream bed*—This characteristic will markedly increase the value of K_d, and is usually found in shallow turbulent streams with a rock bed. The growths would be similar to those found in the trickling filter.
3. *Immediate demand*—Some wastes contain a reducing chemical substance or end-products of decomposition which may cause an immediate oxygen demand.
4. *Nutrients*—If the waste or river is deficient in nitrogen or phosphorus, the oxidation would proceed at a slower rate than that which would be observed in the laboratory, using standard dilution water.

5. *Lag*—If insufficient or inadequate bacteria are present, an interval of time is required for the bacteria to become acclimatized to the waste.

The factors which affect the rate of removal of organic matter, but not necessarily the oxidation rate, are as follows:

1. *Sedimentation and flocculation*—Organic matter in the suspended or colloidal state may settle and decompose anaerobically. This process can produce a marked decrease in the BOD, without an associated change in oxygen levels. The organic fraction of the sludge deposits decomposes anaerobically, except for the thin surface layer which is subjected to aerobic decomposition by virtue of the dissolved oxygen in the overlying waters. In warm weather, when the anaerobic decomposition proceeds at a more rapid rate, gaseous end products, usually carbon dioxide and methane, rise through the supernatant waters. The evolution of the gas bubbles may become sufficiently intense to raise sludge particles to the water surface. Although this phenomenon may occur while the water contains some dissolved oxygen, the more intense action during the summer usually results in the depletion of dissolved oxygen.
2. *Scour*—Organic matter, previously settled, may be picked up by high velocity flow, increasing the BOD values at downstream stations. This factor may cause an apparent decrease in the reaction rate, but significant changes can occur in the oxygen concentration.
3. *Volatilization*—Certain organic compounds may react with end products escaping as gases. As with settling, this factor can result in a significant change in BOD from one station to another without the equivalent oxygen differential.

The effect of these factors is shown graphically in Fig. 4-2. Because of these factors, the rate determined from river data is seldom equal to the rate determined from laboratory data. Thomas (1948) has proposed to combine these effects in another coefficient, K_3, as follows:

$$K \text{ (river)} = K_3 + K_1 \text{ (laboratory)} \tag{4-2}$$

When organic matter (BOD) is removed by any of the above factors,

K_3 is positive; when it is returned to the flowing water and measured at a downstream station, it may be negative. Since K_1 for both the river and laboratory may be readily determined, K_3 can be found by use of Equation (4-2).

The foregoing discussion has dealt with the change in organic matter in a flowing stream, but did not indicate the manner in which

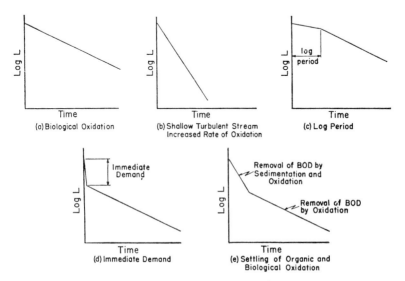

FIG. 4-2. Typical river BOD reactions.

the associated deoxygenation would occur. This rate may be expressed as follows:

$$\frac{dD}{dt} = + K_d L \qquad (4-3)$$

in which

$\dfrac{dD}{dt}$ = rate of change of the dissolved oxygen deficit, and

K_d = coefficient defining deaeration.

In general, when BOD is removed by oxidation only, in any of the fashions indicated above, then it may usually be assumed that the rate of deoxygenation K_d is equal to the rate of BOD removal K_r. On the other hand, if BOD is removed by sedimentation or volatilization, then the measured river rate will be greater than that of oxidation. If longitudinal mixing or scour from high velocity flow is present, it is probable the measured river rate will be less. It is apparent that these factors must be evaluated in order to arrive at representative values of the coefficients of BOD removal and deaeration. The nature and characteristics of the raw waste will usually indicate the possibility of these effects in the river. If a significant portion of the BOD is in suspended form, sedimentation of this portion is possible in the river. BODs run on a sample before and after gas-stripping will indicate the possibility of volatilization. If none of these characteristics is evident from analysis of the raw waste, this point may be confirmed by proper and controlled sampling of the stream. In addition to the standard sampling, which include DO, BOD, temperature and any significant characteristics associated with the waste and river, it is suggested that volatile suspended solids and oxygen uptake rates be measured at all stream stations. The volatile suspended solids indicate settling possibilities of organic matter and the uptake rates may be used as a direct measure of the rate of deaeration. The test is carried out in the following fashion: The river sample is aerated and the drop-off of oxygen is recorded, polarographically. The sample is sealed and no aeration is permitted to occur during this period. Agitation is effected by means of a magnetic stirrer. The time over which this test is conducted depends upon the concentration of the organic matter and its rate of oxidation. The oxygen uptake rate is equal to the slope of the line which defines the oxygen drop-off. It is evident from this description that the uptake rate is equivalent to the rate of BOD test at zero time. These data are particularly valuable when longitudinal mixing, settling and scour are present. Typical values of K_d and K_r for various streams are shown in Table 4-1.

OXYGEN RESOURCES

Dissolved oxygen required for the stabilization of the organic matter is available from three sources: (a) that present in the river and the waste at the point of discharge; (b) that available by means of reaeration and (c) that produced by the photosynthetic activity

TABLE 4-1

River	Flow c.f.s.	Temperature °C	BOD p.p.m.	Reported temperature (K_d) (K_r)		Comments
Elk	5	12	52	3·0	3·0	Shallow rocky bed stream.
Holston	620	22	13	0·15	1·7	Biological and chemical oxidation, and rocky bed.
Wabash	2800	25	14	0·3	0·75	Sedimentation of organic matter.
Wilamette	3800	22	4	0·2	1·0	—

River	Reference
Elk	Camp, Dresser and McKee (1948).
Holston	Kittrell and Kochtitsky (1947).
Wabash	Wabash River-pollution Abatement Needs (1950).
Wilamette	Gleason (1936).

of the green plants. The latter source is usually not relied upon in the determination of the pollution assimilation. However, if significant during a stream survey, it must be taken into account in the analysis of the data. The oxygen available from the first source is readily determined from stream and waste samples. Reaeration is usually the most important factor in the supply of dissolved oxygen. Unpolluted water maintains in solution the maximum quantity of dissolved oxygen, which is in equilibrium with the partial pressure of oxygen in the atmosphere. When oxygen is removed from solution, an unbalance is created and the deficiency is made up by the atmospheric oxygen passing into solution. The rate at which reaeration takes place is proportional to the dissolved oxygen deficit, and may be expressed by Equation 3-10:

$$\frac{dD}{dt} = -K_2 D \qquad (4\text{-}4)$$

D is the dissolved oxygen deficit and K_2 is the reaeration coefficient (base e). A general approximate formula for the reaeration coefficient of natural rivers (O'Connor, 1958) is given by Equation 3-17:

$$K_2 = \frac{(D_L U)^{1/2}}{H^{3/2}} \tag{4-5}$$

in which:

K_2 = reaeration coefficient base e
D_L = diffusivity of oxygen in water
   = 0·000081 ft²/hr at 20°C
U = velocity of flow
H = depth of flow

The effect of temperature on the reaeration coefficient is as follows:

$$K_T = K_{20} \times 1 \cdot 047^{T-20} \tag{4-6}$$

A common range of K_2 is from 0·20 to 10·0 per day, the lower value representing deep slow-moving rivers and the higher value, rapid shallow streams. The depth in Equation (4-5) is equivalent to the ratio of the volume to the surface area of a particular stretch of river or to the ratio of the cross-sectional area to the width at a particular station. The velocity refers to the mean velocity over the cross-section. In order to determine a representative value of the reaeration coefficient, it is necessary to have representative values of depth and velocity. Under steady-state conditions of flow, these values will vary from station to station. The variation depends upon the hydraulic properties of the particular channel: roughness, width and curvature. This variation is subject to statistical analysis and is usually well defined by a normal distribution. Because of this property, it can be readily calculated how closely the measured mean falls within a given percentage of the true mean for any number of measurements. Any value of the reaeration coefficient within limited statistical ranges may be used in oxygen balance calculations. The importance of this factor has been discussed (O'Connor, 1958).

The effect of waste substances in the dissolved or colloidal state can have an appreciable effect on the reaeration coefficient. Surface-

STREAM AND ESTUARY ANALYSIS

active substances tend to concentrate at the air–water interface and apparently create a barrier to the oxygen diffusion. The magnitude of this influence is a function of the type and concentration of the substance. Equally significant is the influence of certain solutions on the solubility of dissolved oxygen. The solubility of oxygen in sea water of various concentrations has been determined and reported and it is probable that many waste substances have similar effects on the solubility. These effects may readily be measured in the laboratory. These factors have been discussed in Chapter 3.

OXYGEN BALANCE IN STREAMS

The simultaneous action of deoxygenation and reaeration produces a pattern in the dissolved oxygen concentration known as "the dissolved oxygen sag". Streeter and Phelps (1925) first presented the theoretical analysis of this relationship in their original work on the Ohio River and developments by others, notably Thomas (1948) and Velz (1939), have improved the utility of this concept. The differential equation describing the combined action of deoxygenation and reaeration is as follows:

$$\frac{dD}{dt} = K_d L - K_2 D \qquad (4\text{-}7)$$

The rate of change in the dissolved oxygen deficit, D, is the result of two independent rates. The first is that of oxygen utilization in the oxidation of organic matter. This reaction increases the dissolved oxygen deficit at a rate that is proportional to the concentration of organic matter, L. The second rate is that of reaeration, which replenishes the oxygen utilized by the first reaction and decreases the deficit. K_d and K_2 are, respectively, the coefficient of deoxygenation and reaeration. Before integrating this equation, the concentration of organic matter, L, must be expressed in terms of the initial ultimate concentration, L_0, at the location of the waste discharge:

$$L = L_0 \, e^{-K_r t} \qquad (4\text{-}8)$$

In this equation, K_r is the coefficient of BOD removal in the stream, which may be significantly different from that of oxidation, K_d.

Substitution of Equation (4-8) for the value, L, in Equation (4-7) and integration yields:

$$D = \frac{K_d L_0}{K_2 - K_r} (e^{-K_r t} - e^{-K_2 t}) + D_0 e^{-K_2 t} \qquad (4\text{-}9)$$

L_0 and D_0 are the initial ultimate biochemical oxygen demand and initial dissolved oxygen deficit, respectively, and refer to the values at the location of the waste discharge. D is the deficit at time, t (Fig. 4-3).

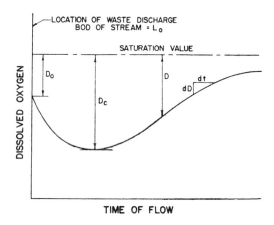

FIG. 4-3. Dissolved oxygen sag curve.

From an engineering viewpoint, the dissolved oxygen sag indicates one point of particular significance; the point of minimum DO concentration or of maximum deficit. This critical point is defined by the numerical equality of the two opposing rates, i.e. the point where the rate of change of the deficit is zero. The balance at this point may be written:

$$K_2 D_c = K_d L = K_d L_0 e^{-K_r t_c} \qquad (4\text{-}10)$$

The time, t_c, to the critical point is:

$$t_c = \frac{1}{K_2 - K_r} \log \frac{K_2}{K_r} \left(1 - \frac{D_0(K_2 - K_r)}{K_d L_0} \right) \qquad (4\text{-}11)$$

The above equations are similar to those initially presented by Streeter and Phelps. However, the difference between BOD removal (K_r) and oxygen utilization by the organic matter (K_d) has been taken into account. Inspection of Equations (4-10) and (4-11) indicates that the allowable pollutional load L_0 that the stream may absorb is a function of: (a) the allowable dissolved oxygen deficit, D_c; (b) the coefficient of deoxygenation, K_d, and reaeration, K_2, and of BOD removal, K_r; (c) the initial deficit, D_0. The first factor is usually established by standards of the health agency and the last by local conditions resulting from upstream pollution. The engineering problem is usually associated with the assignment of representative values of the coefficients. Field data from controlled stream surveys are required for the determination of these coefficients.

In the mathematical development of these equations it is assumed that K_r, K_d and K_2 are constant, that only one source of pollution exists and that there is no tributary inflow or pollution. Variations from these assumptions may be taken into account in any practical case.

Stream Surveys

From the above discussions, it may be seen that data from controlled stream surveys are required for the determination of the coefficients. The survey should be conducted as much as is practical and possible under a steady-state condition of river flow, waste flow and temperature. Cross-sectional measurements are necessary for the determination of the depths. With this information the velocity may be calculated from the continuity equation or it may be measured directly. Both depth and velocity are required for the reaeration coefficient and the velocity is also used to calculate the time of flow between stations.

Ordinarily biochemical oxygen demand, dissolved oxygen, suspended solids, temperature and pH are the minimum requirements for both river and waste sampling. On a few of the BOD samples, long-term determinations should be made which are used in the computation of k_1. Oxygen uptake curves should be run at all stations. If settling is a possibility, volatile suspended solids should be included in the sampling as these data will indicate what portion

Analysis of Data

Calculation of the stream rate of removal, K_r, is made in accordance with Equation (4-1), using the BOD data from the respective stream stations. K_1 is determined from the long-term laboratory data. Assignment of K_d must now be made in accordance with the above discussions. Equation (4-5) permits determination of K_2.

Knowing the initial BOD and oxygen deficit, the dissolved oxygen sag curve (Equation 4-9) may be computed and compared to the observed values. Adjustment of the coefficients, within statistical limits defined by the data, may be necessary to secure the best fit between the calculated and observed points. Having established coefficients for one set of flow and temperature conditions, they may be extrapolated to any other set of appropriate conditions. Consideration must be given to possible changes in the deaeration and removal coefficients due to anticipated treatment.

The fundamental principles, as expressed by the above equations, may now be employed in the evaluation of the natural purification of a stream. Assuming knowledge of the rates of deoxygenation and reaeration in the receiving stream, the first important relationship to be developed is that of the dissolved oxygen concentration at the critical point for various river flows, for a given waste loading. The waste loading, which is used in developing this relationship, should be the average BOD of waste discharged from the plant. This relationship indicates in a quantitative manner the effect of the present pollutional load on the dissolved oxygen concentration in the river. This may further indicate the need for waste treatment, in order to maintain standards established for the receiving stream. One of the common and logical standards applies to a minimum allowable concentration of dissolved oxygen. For this minimum concentration, the allowable waste loading may be calculated for various river flows in accordance with Equations (4-10) and (4-11). This analysis will indicate the possible need and the required degree of waste treatment and may be used to establish a probably seasonal plan of operation for the waste treatment plant.

Churchill (1954) has suggested a statistical analysis of stream data in order to determine pollution capacity without using the stream coefficients, as described above.

Minimum Stream Flow

For a given waste discharge, the dissolved oxygen sag curve will vary in time and magnitude in accordance with the river flow. In this regard, some minimum stream flow is of particular significance. In some cases, the receiving stream dries up completely, the only flow being that of the discharged waste. In the other cases, where some flow occurs even during the driest period of the year, the problem of selection of a minimum flow is present. In selecting the minimum flow upon which to base a stream analysis, consideration should be given to the following factors:

(1) Stream usage.
(2) Probability of minimum flow and its duration.
(3) Damage if flow less than minimum occurs.
(4) Cost of increased treatment to meet more stringent requirements.

Although most of the probability studies are based on daily values, it has been suggested that in some cases minimum weekly or minimum monthly flows be used. In certain areas, where stream beds may be dry for periods of weeks and even months, statistical analyses are not appropriate. Applicable in this case, is the duration curve which is a tabulation or plot of river flows against total occurrences not exceeded by these flows.

OXYGEN BALANCE IN ESTUARIES

The forces of self-purification in tidal bodies are similar to those in flowing streams. All the factors which determine the oxygen balance in rivers are likewise active in estuaries. The one important difference, however, is the turbulent diffusion caused by the tidal action and its effect on the time of travel of an average element of pollution. In a flowing stream, the predominant motion is due to gravity and the basic formulations describing this motion are sufficiently accurate to permit calculation of the time of travel. The time of travel is markedly changed, however, in those sections of the river which are subjected to tides. The motion of these waters is caused not only by gravity but also by tidal action, density currents, and wind effects. Waste materials which are discharged into estuaries are mixed with the water and are gradually diminished in concentration by the tidal motion, which carries various portions of the pollutant back and forth over many cycles. Any slug of

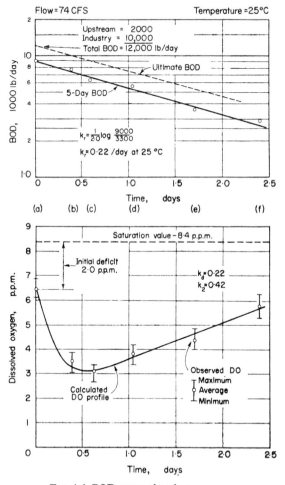

FIG. 4-4. BOD removal and oxygen sag.

Figures 4-4 and 4-5 are referred to on page 146.

pollutant is ultimately translated to the open sea and this time is referred to as the flushing time rather than the travel time. The flushing time is the resultant of the translating velocity of the river flow and the longitudinal mixing of the tidal action. This concept is of fundamental importance in defining pollution and self-purification of tidal waters.

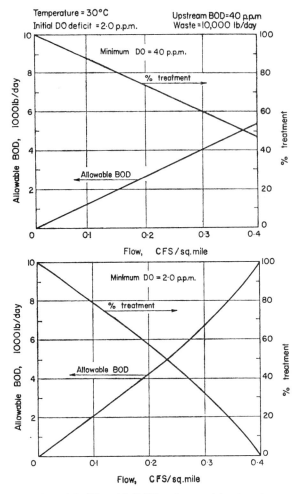

FIG. 4-5. Allowable BOD and percent treatment.

Figures 4-4 and 4-5 are referred to on page 146.

The mechanism of estuarine mixing and flushing is not yet completely generalized in a theoretical manner. Although the state of knowledge of these phenomena is far from complete, it has progressed to the point where many useful concepts are currently available to the engineer. A review of estuarine hydrography has been presented by Pritchard (1952) and a study of the dissolved oxygen profiles in estuaries by O'Connor (1960).

In the following discussion, non-stratification in both the vertical and lateral planes is assumed. Consider an estuary receiving no other waste matter along its course except that contributed by the land runoff from an upstream discharge. A steady-state condition is assumed to exist which implies the land runoff and the waste discharge have remained constant for a period equivalent to the time of passage through the estuary. The deaeration caused by the organic matter and the resulting reaeration produce a pattern in the dissolved oxygen concentration similar to the "sag curve" evidenced in upland rivers. Dissolved oxygen profiles are shown in Fig. 4-6, indicating the average condition and those at the flood and ebb slacks. The latter two curves delineate the limits between which the dissolved oxygen profiles move over the tidal cycle. The longitudinal displacement between these limits depends upon the tidal velocity and period and the turbulence due to tidal motion. The displacement is a maximum for large values of tidal velocity and period and relatively low order turbulence. Conversely, intensely mixed tidal bodies of relatively low velocity and short period are characterized by a minimum displacement of the profiles. The cyclical variation of dissolved oxygen in time is readily defined from these profiles. The dissolved oxygen varies from a low to a high value over the tidal cycle as shown in Fig. 4-6. In regions where the longitudinal concentration gradient is great, there is a great variation in the dissolved oxygen over the cycle, as at stations 1 and 5. Where the concentration gradient is zero or approaches zero, the variation over the cycle is low, as at stations 3 and 6. Where the concentration gradient is negative, the dissolved oxygen varies inversely as stage height, as at station 1 and where positive, as at station 5, the variation is direct. Inspection of Fig. 4-6 indicates that the dissolved oxygen concentration remains substantially constant in time at the flood and ebb slack periods. The profiles therefore can be more readily defined and more accurately measured at these periods than any other.

The differential equation describing the oxygen balance in an estuary is similar to that for rivers,

$$\frac{dD}{dt} = U\frac{dD}{dx} = K_d L - K_2 D + E_x \frac{d^2 D}{dx^2} \qquad (4\text{-}12)$$

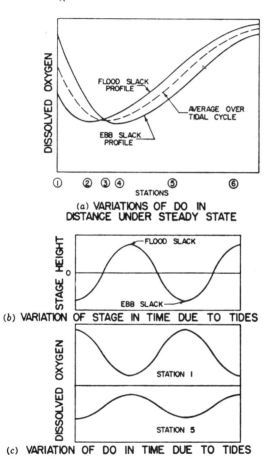

FIG. 4-6. DO variation due to tides.

E_x is the turbulent diffusion coefficient. In addition to the terms of the river oxygen sag, equation (4-12) also includes the term defining the turbulent mixing, which is comparable to the longitudinal mixing to rivers. The magnitude of this mixing effect in estuaries is of the order of 10 to 100 times that in rivers. Before Equation (4-12) can be integrated, the variable L must be expressed as a function of x. This may be done by a material balance similar to that used in the development of the oxygen balance. It may be shown that:

$$L = L_0 e^{J_1 x} \qquad (4\text{-}13)$$

in which

L = BOD at a distance x downstream from the source of pollution

L_0 = BOD at the source

$$J_1 = \frac{U}{2E}\left[1 - \sqrt{\left(1 + \frac{4K_d E}{U^2}\right)}\right]$$

Substitution of Equation (4-13) in Equation (4-12) for the value of L and integration leads to:

$$D = \frac{K_d L_0}{K_2 - K_d}[e^{J_1 x} - e^{J_2 x}] + D_0 e^{J_2 x} \qquad (4\text{-}14)$$

in which

$$J_2 = \frac{U}{2E}\left[1 - \sqrt{\left(1 + \frac{4K_2 E}{U^2}\right)}\right]$$

If the land runoff is very low or the cross-sectional area of the estuary very large, the velocity term becomes insignificant. For this case,

$$J_1 = \left[\frac{K_d}{E}\right]^{1/2}$$

$$J_2 = \left[\frac{K_2}{E}\right]^{1/2}$$

Before Equation (4-14) can be employed in a practical case, the velocity term, the eddy diffusivity and the coefficients of reaeration and deaeration must be assigned numerical values.

Evaluation of the Coefficients

The turbulent transport coefficient may be determined from a knowledge of the concentration distribution of any property in the estuary. The chloride concentration of the salinity is taken as the required steady-state property. The source of this property is the coastal waters of the ocean. The turbulent flux, caused by the tidal action, diffuses salt waters into the mouth of the river. The distribution and extent of this characteristic reflects the magnitude of the eddy transport coefficient. The steady-state condition predicates fresh water from one major source, with none added by ground water and removed by evaporation. The material balance of this property may be constructed in a fashion similar to that of the oxygen-demand material.

Assuming constant values of the eddy diffusivity and velocity terms, integration of the chloride balance equations yields:

$$S = S_0 \, e^{-(U/E)x} \qquad (4\text{-}15)$$

in which
S = salinity at distance, x
S_0 = salinity at $x = 0$

A plot of the logarithm of the saline or chloride concentration against an arithmetic time scale will indicate a linear relation if the coefficients U and E do not vary with distance. The slope of such a line is a measure of the ratio of these coefficients and E may be evaluated graphically in this fashion. In some cases, the section of river under study may be above the intrusion of sea water and salinity data would not be available for evaluation of the turbulent transfer coefficient. In this case, any conservative property of the waste water, which significantly increases the river water concentration, may be used to determine the eddy coefficient.

A particularly appropriate characteristic is the concentration of a stable dissolved chemical. The following equation is appropriate for this case:

$$\frac{S}{S_0} = \frac{1 - e^{(U/E)x}}{1 - e^{(U/E)x_0}} \qquad (4\text{-}16)$$

in which

S_0 = upstream concentration of the conservative property
S = downstream concentration of the conservative property
x_0 = distance from the mouth of the estuary or from the point where the concentration becomes constant to the point where concentration = S_0
x = distance from the mouth of the estuary or from the point where the concentration becomes constant to the point where concentration = S, the distance increasing from the mouth or from the point of constant concentration.

The flow velocity, U, may be calculated directly from the continuity equation:

$$U = \frac{Q}{A} \qquad (4\text{-}17)$$

in which

Q = river discharge
A = cross-sectional area

It is probable that the velocity will also vary with distance, decreasing in the downstream direction. The error introduced by assuming an average value may be determined as in the case of the eddy diffusivity. Frequently, the value of this velocity is so small that no significant error is introduced in the oxygen calculations by this assumption. In any case, the cross-sectional area is frequently composed of two distinct areas: the relatively shallow littoral zone, and the deep central channel zone. The quantity of flow in this latter zone is usually so much greater than in the littoral area that it may be appropriate in determining the velocity by means of Equation (4-17), to employ only the central channel area. Each estuary must be evaluated separately in this regard and examination of the shape of the cross-sectional areas indicates possibility of this channeling effect.

The reaeration coefficient may be calculated by means of Equation (4-18). This equation was developed on the basis that both the vertical velocity fluctuation and the mixing length are approximately equal to one-tenth of the forward flow velocity and of the

STREAM AND ESTUARY ANALYSIS

average depth, respectively. It is significant to note that these ratios are of the same order of magnitude as in non-tidal rivers. Substituting the ratio of the average tidal velocity and average depth for the rate of surface renewal, the reaeration coefficient becomes:

$$K_2 = \frac{(D_L U_0)^{1/2}}{H^{3/2}} \tag{4-18}$$

The velocity U_0 refers to mean tidal velocity and not the land runoff velocity. The depth refers to the average over the section and over a cycle. If both the tidal velocity and the depth vary considerably with distance, then short stretches must be taken over which the variation is minimal.

The evaluation of the rate of deoxygenation may be carried out in a manner similar to that described for rivers, allowing for the eddy diffusivity. Assuming representative values of the eddy diffusivity and the velocity are available, the value of the deoxygenation coefficient, K_d, may be obtained from Equation (4-13). The logarithm of the concentration of organic matter (BOD) is linear with distance for the stated conditions. The slope of the line defining this relation is a measure of the exponent in Equation (4-13) from which K_d may be calculated. In assigning a value of K_d to represent the decay of organic matter, distinction should be made between the removal of organic matter and the rate of deoxygenation as described above.

Examples of the calculated dissolved oxygen profiles and the associated BOD and chloride curves for the estuaries of the Delaware River is shown in Fig. 4-7. In the case of the Delaware River sampling was made during the slack periods (Kaplovsky, 1957). The coefficients are calculated in accordance with the procedures outlined above.

Minimum DO and Allowable BOD

The co-ordinates of the point of minimum dissolved oxygen are as follows:

$$x_c = \frac{1}{J_1 - J_2} \log \frac{J_2}{J_1} \left[1 - \frac{D_0}{L_0 F} \right] \tag{4-19}$$

$$D_0 = \frac{K_d L_0}{K_2} e^{J_1 x} \left[1 + \frac{J_1^2 E}{K_2 - K_d} \right] - \frac{E J_2^2 e^{J_2 x}}{K_2} [FL_0 - D_0] \tag{4-20}$$

144 BIOLOGICAL WASTE TREATMENT

From these equations, the allowable BOD loading L_0 may be determined for any set of conditions of land runoff, temperature and turbulence.

Example 4-1

Assume an industry is presently discharging 7500 lb/day of 5-day BOD. The ultimate BOD is 10,000 lb/day. Stream standards stipulate

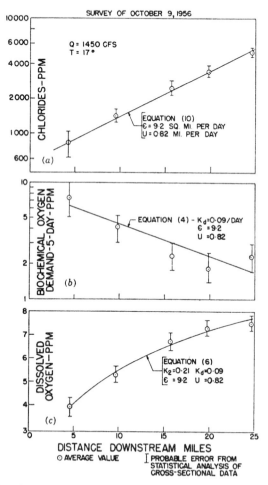

FIG. 4-7. Chloride, BOD, and DO profiles for the Delaware River

that the minimum dissolved oxygen in the river shall not fall below 4 p.p.m. under ordinary low-flow (0·25 ft^3/sec/mile^2) and 2 p.p.m. under extreme low flows (0·1 ft^3/sec/mile^2). It is necessary to determine the degree of treatment required under these conditions for the plant expansion.

A stream survey was conducted at a flow of 0·4 ft^3/sec/mile^2 and the pertinent data of BOD and DO are shown in Table 4-2. Survey data also include suspended solids, per cent volatile, pH and immediate oxygen demands. From analysis of these data and inspection of the river, neither settling nor toxic effects nor immediate demands is significant. The average values of depth and velocity over the stretch of river surveyed are also indicated. From topographic maps, distances between stations and drainage areas tributary to stations were measured. Long-term BOD runs on river and waste samples were carried on in the laboratory and indicated that laboratory coefficient, k_1, was constant for all stations with a value of 0·12 per day at 20°C. Since the ratio of the 5-day BOD to the ultimate BOD remains constant, either set of values may be used in the determination of the removal coefficient.

TABLE 4-2. SURVEY DATA

Station	Average BOD–5 day p.p.m.	Average DO p.p.m.	Mile point	Drainage area miles^2
A	22·5	6·4	39·0	186
B	16·3	3·5	34·5	217
C	11·4	3·1	32·0	256
D	8·0	3·8	26·9	322
E	4·3	4·3	19·1	395
F	3·0	5·8	11·0	458

1. Hydrometric data of survey:
   Flow   = 0·4 ft^3/sec/mile^2
   Velocity   = 0·71 ft/sec
   Depth   = 5·9 ft
   Temperature = 25°C.

2. Laboratory long-term BOD runs of river samples show:
   k_1 = 0·12 per day at 20°
   BOD 5-day = 0·75 BOD ultimate.

The first step in the calculation is the determination of the coefficient of BOD removal. The time of flow is calculated from the velocity and distance data and the cumulative time is shown in Table 4-3. No significant tributaries enter the main stream. The tributary flow is uniformly distributed along the stream and the flow at each station therefore increases in direction proportion to the drainage area. These flows are indicated in Table 4-3. It is assumed that the tributary flow is fully saturated and is free of organic matter. In order to account for the dilution effect, the BOD at each station is expressed in pounds per day. These values are shown in Table 4-3. A plot of the logarithms of the BOD against a linear scale of time is shown in Fig. 4-4. A line of best fit by eye is constructed and the slope of this line is the value of the coefficient of BOD removal, k_r, at the temperature of the stream survey. A typical calculation using Equation (4-1) is shown on the figure for a time interval of two days. Since there is no evidence to indicate BOD removal other than by oxidation, it is assumed the $k_r = k_d$ in this case.

The second step is the calculation of the reaeration coefficient in accordance with Equation (4-5). The pertinent calculation and units are shown in Table (4-3). Since the calculation is made on the basis of 20°C, Equation (4-6) must be employed to convert this value to the stream temperature of 25°C.

The third step is the determination of the oxygen sag curve in accordance with Equation (4-9). A sample calculation is shown below between stations A and B:

Station A

BOD 5-day   = 7500 (industry) + 1500 (upstream)

                      = 9000 lb/day

L_0 = 12,000 lb/day at Station A, the ultimate BOD of both waste and upstream pollution

STREAM AND ESTUARY ANALYSIS

TABLE 4-3. CALCULATION OF DEOXYGENATION COEFFICIENT

Station	1 Cumulative time of travel days	2 Flow Mgal/day	3 BOD 5-day lb/day
A	0	48	9000
B	0·39	56	7610
C	0·60	66	6310
D	1·04	83	5540
E	1·71	102	3650
F	2·40	118	2950

1. Calculated from data of velocity and distance.
2. Calculated from 0·4 ft^3/sec/mile^2 and drainage area at station.
3. lb/day = BOD × flow × 8·34.

CALCULATION OF REAERATION COEFFICIENT

U = 0·7 ft/sec = 2520 ft/hr
H = 5·9 ft
D_L = 0·000081 ft^2/hr at 20°C

$$K_2 = \frac{[0 \cdot 000081 \times 2520]^{1/2} \, 24}{5 \cdot 9^{3/2} \cdot 2 \cdot 30} = 0 \cdot 33 \text{ per day at } 20°C$$

$K_{25} = K_{20} \cdot 1 \cdot 047^{(t-20)} = 0.38$ per day at 25°C.

$Q = 0 \cdot 4 \text{ ft}^3/\text{sec/mile}^2 \times 186 = 74 \cdot 5 \text{ ft}^3/\text{sec} = 48 \text{ Mgal/day}$

Temperature = 25°C

Saturation = 8·4 p.p.m.

Dissolved Oxygen = 6·4 p.p.m.

Deficit = 2·0 p.p.m.

= 2·0 × 8·34 × 48 = 800 lb/day

t = Time of flow between A and B = 0·39 days

148 BIOLOGICAL WASTE TREATMENT

k_r $= k_d = 0.22$ per day

k_2 $= 0.42$ per day

Station B

$$D = \frac{0.22 \times 12,000}{0.42 - 0.22}$$

$$[10^{-0.22 \times 0.39} - 10^{-0.42 \times 0.39}] + 800 \times 10^{0.42 \times 0.39}$$

$D = 1810 + 580 = 2390$ lb/day

Flow at B $= 56$ Mgal/day

$$D = \frac{2590}{8.34 \times 56} = 5.1 \text{ p.p.m.}$$

DO $= 8.4 - 5.1 = 3.3$

Similar computations are performed for the remaining stations and the calculated curve is shown in Fig. 4-4. The agreement between the observed and calculated values indicate that representative values k_r, k_d and k_2 have been selected.

The final step is the determination of allowable BOD and required per cent of treatment for the various conditions as outlined initially. Assuming a critical temperature of 30°C during the drought period and a minimum allowable DO of 4 p.p.m. at a flow of 0·25 ft³/sec/mile², a sample calculation proceeds as follows:

$$D_c = 7.6 - 4.0$$

$$= 3.6 \text{ p.p.m.}$$

$$= \text{allowable DO deficit at minimum point.}$$

For a flow of 0·25 ft³/sec/mile², the average depth in the river is 5·4 ft and the average velocity 0·63 ft/sec. At a temperature of 30°C using Equations (4-5) and (4-6).

$$k_2 = 0.56 \text{ per day}$$

The BOD coefficient must also be converted to 30° temperature by Equation (4-6)

$$k_d = k_r = 0.28 \text{ per day}$$

The time of flow to the critical point, t_c, must now be determined. Inspection of Equation (4-11) indicates that a value of the allowable BOD, L_0, must be assumed. The assumed value is then compared to the value calculated from Equation (4-10). Trial and error is necessary but in many cases the ratio of the initial deficit, D_0, to the allowable loading, L_0, is small and therefore not significant. If an average initial deficit is taken as 2·0 p.p.m. (as determined from survey data) and the ratio of D_0/L_0 assumed to be 0·10 ($L_0 = 20$ p.p.m.) then the critical time is determined as:

$$t_c = \frac{1}{0.56 - 0.28} \times \log \frac{0.56}{0.28} \left[1 - 0.10 \frac{(0.56 - 0.28)}{0.28} \right]$$

$$t_c = 0.91 \text{ days}$$

Knowing the velocity of flow at this condition, the location of and the drainage area tributary to the critical point may be determined. For the above conditions, the drainage area is about 260 miles square at this point, and the flow 42 Mgal/day. For the allowable deficit of 3·6 p.p.m.

$$D_c = 42 \times 3.6 \times 8.34 = 1260 \text{ lb/day}$$

This value is substituted in Equation (4-10) with the appropriate values of k_2, k_d and t_c to solve for the allowable pollutional load.

$$L_0 = 1260 \times \frac{0.56}{0.28} \times 10^{+0.28 \times 0.91}$$

$$= 4550 \text{ lb/day}$$

The flow at the initial point is 30 Mgal/day and the BOD is therefore 18 p.p.m., which is close enough to the assumed value of 20 p.p.m. Similar values, computed for various runoffs, are shown in Fig. (4-5). In order to determine the required per cent treatment, upstream pollution must be subtracted from the total allowable load. Assuming 4·5 p.p.m. as an average BOD residual from upstream, the allowable load from the industry is:

$$L_0 = 4550 - 1130$$

$$= 3420 \text{ lb/day}$$

The 5-Day industrial BOD converted to its ultimate value = 10,000 lb/day. The required per cent treatment is therefore:

$$\% = \frac{10,000 - 3420}{10,000} = 66\%$$

A comparable set of calculations is performed for a minimum DO of 2·0 p.p.m. The allowable BOD and per cent treatment for both 2·0 and 4·0 p.p.m. minimum DO are shown in Fig. 4-5.

REFERENCES

1. CAMP, DRESSER and McKEE, Sanitary Water Board, Department of Health, Commonwealth of Pennsylvania (March, 1949).
2. CHURCHILL, M. A., *Sew. and Ind. Wastes*, **26**, No. 7, 887 (July 1954).
3. GLEASON, G. W., *Oregon State Agricultural College Bulletin*, Series No. **6** (April 1936).
4. KAPLOVSKY, A. J., *Sew. and Ind. Works*, **29**, No. 9, 1942 (Sept. 1957).
5. KITTRELL, F. W. and KOCHTITSKY, O. W., *Sew. Works J.*, **19**, No. 6 (Nov. 1947).
6. NUSBAUM, I. and MILLER, H. E., *Sew. & Ind. Wastes J.*, **24**, No. 12 (Dec. 1952).
7. O'CONNOR, D. J., *Proc. Amer. Soc. Civil Engrs.*, S.E.D. (1960).

8. O'CONNOR, D. J. and DOBBINS, W. E., *Trans. Amer. Soc. Civil Engrs.*, **123**, 641 (1958).
9. O'CONNOR, D. J., "Oxygen Relationships in Streams": Robert A. Taft Sanitary Engineering Center, Technical Report W 58-2, Cincinnati, Ohio (1958).
10. PRITCHARD, D. W., Technical Report VII, The Chesapeake Bay Institute, Johns Hopkins University, (1954).
11. STREETER, N. W. and PHELPS, E. B., *Public Health Bull.* 146, USPHS (1925).
12. STOMMEL, H., *Sew. and Ind. Wastes J.*, 25 (Sept. 1953).
13. THOMAS, H. A., Jr., *Water and Sew. Works*, **95**, No. 9 (1948).
14. VELZ, C. J., *Trans. Amer. Soc. Civil Engrs*, **104** (1939).
15. "Wabash River-Pollution Abatement Needs", Ohio River Valley Water Sanitation Commission (Aug. 1950).

CHAPTER 5

SOLID–LIQUID SEPARATION

SEDIMENTATION

SEDIMENTATION is a waste treatment process whereby suspended and coagulated particles of density greater than that of the liquid medium are removed. Sedimentation may be classified as primary or secondary. The purpose of primary settling is to reduce the suspended and organic load on subsequent treatment units. Where required BOD removal is low and where a significant portion of the BOD is in suspended form secondary treatment processes may not be required. Secondary sedimentation is employed for the clarification of sludge waste mixtures and for the thickening of biological sludges.

Sedimentation has been divided into four classifications depending upon the concentration of the suspension and the coagulating tendency of the particles (Fitch, 1958). These are discrete settling, flocculent settling, zone settling and compression. In the first type, a particle maintains its individuality and does not change in size, shape or density. Flocculent settling is characterized by agglomeration of the particles which is associated with a changing settling rate. In zone settling the particles settle as a mass and exhibit a distinct interface between the supernatant and the settling solids. During compression, solids are mechanically pressing on layers beneath, resulting in a slow displacement of liquid.

Discrete and Flocculent Settling

Under quiescent conditions, the factors which influence the settling process are the size, shape and density of the suspended particles and the viscosity and density of the liquid medium. The viscosity in turn is a function of the liquid temperature. From these characteristics the settling velocity of a particle may be defined. If this velocity remains constant during the settling process, the particle is described as discrete: settling may be described by either a laminar or turbulent flow condition. The settling velocity of a

SOLID–LIQUID SEPARATION 153

spherical particle for both these flow regimes generally is defined by Newton's law.

$$V = \frac{4g\,(\rho_s - \rho_L)\,D}{3\,C_d \rho_L} \qquad (5\text{-}1)$$

$P_S =$ specific gravity of the particle

$P_L =$ specific gravity of the liquid

$D =$ particle diameter

C_d is the drag coefficient which is a function of the Reynolds Number (R_N). When R_N is small, which implies small spheres and low velocity, viscosity is the predominant force and

$$C_d = 24/R_N \qquad (5\text{-}2)$$

This defines the laminar settling regime and the settling velocity for this condition as developed by Stokes is:

$$V = \frac{1}{18}\frac{(\rho_s - \rho_L)}{\mu}\,gD^2 \qquad (5\text{-}3)$$

$\mu =$ the liquid viscosity

Temperature is an important factor in this case, due to its influence on viscosity. As R_N increases a transition zone between laminar and turbulent settling is encountered in which both the viscous and inertia forces are significant. This zone is generally defined by a range R_N of 2–500 in which

$$C_d = \frac{18 \cdot 5}{R^{0 \cdot 6}} \qquad (5\text{-}4)$$

As R_N increases above 500, the inertia force is the predominant factor implying large spheres and high settling velocities. The drag coefficient is constant at 0·4 and temperature is no longer significant and the settling velocity is defined by Equation (5-1). Figure 5-1 relates the settling velocity with diameter for discrete settling.

The application of this theory to the design of settling tanks was first presented by Hazen (1904). His analysis, which applied specifically to discrete particles, took into account short-circuiting. This phenomenon may be defined as the unequal times of passage of

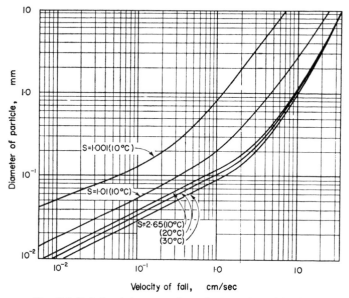

FIG. 5-1. Relation between settling velocity and particle diameter. (from Kalinske, 1946).

fluid through the tank. Camp (1946) developed this analysis and showed the significance of scour and turbulence. The assumptions which are the basis of these analyses are as follows:

(1) The particles are uniformly distributed over an influent cross-section of the tank.

(2) A particle is removed when it hits the bottom of the tank on the sludge zone.

The overflow rate may be defined as the settling velocity, V_0, of a particle that settles through a distance exactly equal to the effective depth of the tank during the theoretical detention period. The relationship between the settling velocity or overflow rate and the dimensions of the settling tank is as follows:

$$A_s = \frac{Q}{V_0} \tag{5-5}$$

in which

A_s = tank surface area

Q = the waste flow

V_0 = average settling velocity

All particles with settling velocities greater than V_0 will be completely removed and particles with settling velocities less than the overflow rate will be removed in the ratio V/V_0. These removal conditions apply provided the flow-through velocity does not exceed the scour velocity of any particles which are to be removed. The scour velocity is defined as follows:

$$V_c = \sqrt{\frac{8\beta}{f} gD(S-1)} \tag{5-6}$$

in which

V_c = velocity of scour

β = constant $\begin{cases} 0.04 \text{ for unigranular sand} \\ 0.06 \text{ for non-uniform sticky material} \end{cases}$

f = Weisbach D'Arcy friction factor (0·03 for concrete)

D = particle diameter

S = specific gravity

Example 5-1. Given a suspension of sand in water with a particle size of 0·088 mm. If the suspended solids content is 400 p.p.m. in a flow of 0·8 Mgal/day at a temperature of 20°C, design a settling tank to remove 75 per cent of the particles. Determine the flow velocity such that all particles of lower settling velocity than those to be completely removed will be scoured.

for $D = 0.088$ mm; $V = 0.65$ cm/sec $= 13,800$ gal/day/ft²

$$\text{S.A.} = \frac{800,000}{13,800} = 58 \text{ ft}^2 \text{ for 100 per cent removal}$$

For 75 per cent removal $A_s = 0.75 \times 58 = 43.5$ ft²

$$V_c = \sqrt{\frac{8\beta}{f} gD(S-1)}$$

$$\left[8 \times \frac{0.04}{0.03} \times 9800 \ (2.65 - 1)(0.088)^{1/2}\right]$$

$$= 125 \text{ mm/sec}$$

$$= 0.4 \text{ ft/sec}$$

$$\text{Cross-sectional area} = A_c = \frac{0.8 \times 1.55}{0.4} = 3.1 \text{ ft}^2$$

Any practical combination of length, width, and depth may be used to meet these requirements.

When the suspended solids are discrete, a graphic representation of a settling path in an ideal tank will yield a straight line, as shown on Fig. 5-2. When flocculation occurs, the settling velocity of the particle is increased as it settles through the tank depth due to coalescence with other particles, producing the curvilinear path, as shown in Fig. 5-2.

Camp (1953) considered that flocculation was due to (a) differences in the settling velocities of the particles, where rapid settling particles overtake slower settling particles and coalesce with them, and (b) velocity gradients whereby particles in regions of higher velocity overtake and coalesce with particles in regions of lower velocity. It was further shown that in relatively deep tanks of low flow-through, flocculation is primarily due to the effect of differential settling velocities.

The suspended solids, encountered in industrial wastes and domestic sewages, are usually of a flocculent nature. For a discrete suspension, the efficiency of removal depends only on the settling velocity or overflow rate, while, for a flocculent suspension, the efficiency is a function of both the settling velocity and the detention time. In the latter case, there is no satisfactory method available for evaluating the flocculation effect and it is therefore advisable to

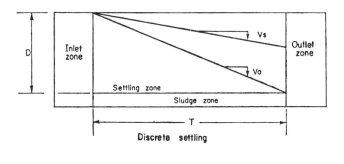

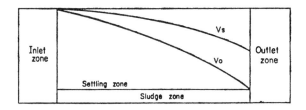

Flocculent settling
FIG. 5-2. Settling paths in an ideal tank.

conduct laboratory experiments in order to measure these factors (O'Connor and Eckenfelder, 1958). With this information it is possible to establish design criteria for settling facilities.

Laboratory Studies

The laboratory study consists of measuring the concentration of suspended solids at various depths and time intervals in a test cylinder. The concentration of the suspended solids is uniform throughout the depth at the start of the test. The depth of the test cylinder should approximate to the effective settling depth of the prototype tank and a constant temperature should be maintained throughout the test period. A practical depth of test cylinder is 8 ft, with taps at depths of 2, 4, 6 and 8 ft. Samples are drawn off at selected time intervals, depending upon the nature of the waste up to 120 min. Suspended solids determinations are made on all samples. The data collected at the 2, 4 and 6 ft depths are employed to determine the settling velocity and time relationships and the data from the 8 ft tap for sludge concentration and compaction. This unit is shown in Fig. 5-3.

The concentrations of suspended solids measured at each tap and time are expressed as a per cent of the initial concentration. The difference between this value and 100 per cent is a measure of the fraction of particles which have settled past this point in the test cylinder. These differences are then plotted against their respective depths and times, as shown in Fig. 5-4. Smooth curves are then approximated, connecting points of equal concentration. The curves shown in this figure represent the limiting or maximum settling paths for the indicated per cent, i.e. the specified per cent of the suspended solids have a settling path equal to greater than that shown and would therefore be removed in an ideal settling tank of the same depth and detention time. The curvilinear paths are indicative of the flocculating nature of the waste. The more pronounced the curvature of the settling paths, the greater the flocculating effect. It is emphasized that currents, induced by thermal effects, will alter the settling velocities of particles. In some cases, slight temperature differences will result in apparently erratic settling curves. Domestic sewage appears to be particularly subject to this influence.

Analysis of Settling Data

From the relationships as shown on Fig. 5-4 the final design curves may be developed in a manner similar to that described by Camp (1946) for discrete settling. An effective settling velocity, V_0, may be defined as the effective depth, 6 ft, divided by the time required for a given per cent to settle this distance. All particles having a settling velocity equal to or greater than V_0 will be removed in an ideal settling basin having an overflow rate of V_0. Particles

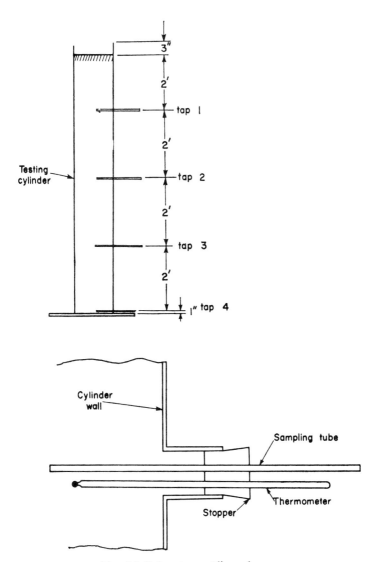

Fig. 5-3. Laboratory settling column.

with a lesser settling velocity, V_s, will be removed in the proportion V_s/V_0. For example, referring to Fig. 5-4 with a detention time of 25 min and a 6 ft settling depth ($V_0 = 14.4$ ft/hr), 50 per cent of the suspended solids are completely removed—i.e. 50 per cent of the particles have settling paths equal to or steeper than that shown. Particles in the next 10 per cent range (50 per cent–60 per cent) will

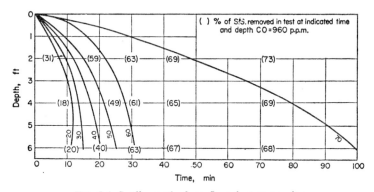

FIG. 5-4. Settling paths for a flocculent suspension.

be removed in proportion to the ratio V_s/V_0. The average depth to which this range will settle is 3·8 ft, as shown on Fig. 5-4. The associated settling velocity is 3·8 ft divided by 25 min, which is equal to 9·1 ft/hr. Each subsequent percent range is computed in a similar manner and the total removal is developed as follows:

%S.S. range	V_s ft/hr	V_s/V_0	%S.S. removed
0–50	14·4	1·0	50
50–60	9·1	0·63	6·3
60–70	3·6	0·25	2·5
		Total removed	58·8

The total removal of the suspended solids is effected under ideal settling conditions of 25 min with a 6 ft depth which is equivalent to an overflow rate of 2600 gal/day/ft². In a similar fashion, various per cent removals and their associated detention times and overflow rates may be computed.

Since the flocculating effects of any waste apparently varies as the initial concentration, a series of similar analyses should be conducted for the anticipated range of suspended solids. From these analyses a range of per cent suspended solids removal with respect to overflow

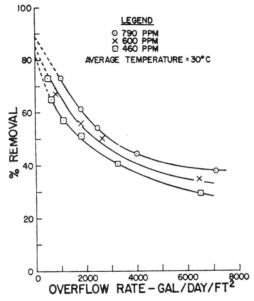

FIG. 5-5. Suspended solids removal vs. overflow rate.

rate and detention time can be established as shown on Figs. 5-5 and 5-6.

In many wastes, a fraction of the suspended solids is not removable by sedimentation. The curves developed from the laboratory analysis (Figs. 5-5 and 5-6) approach the removable fraction as a limit. The removal relationships are influenced by the nature of the waste and the initial concentration of suspended solids. The per cent removal increases with increasing concentration. This effect is shown on Figs. 5-5 and 5-6. The nature of the waste may have a significant effect on the removal characteristics. For example, a pulp mill waste, on occasions, may contain lime, resulting in a significant increase in removal efficiency as contrasted to this same waste with no lime present. This is attributed to both the flocculating effect of the lime and the greater density of the suspension.

Design Criteria

In order to develop criteria for prototype design, factors in addition to the laboratory settling velocities, detention times and sludge

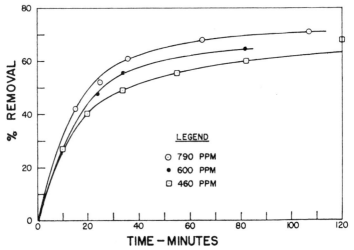

FIG. 5-6. Suspended solids removal vs. detention period.

compaction must be considered. The efficiency of the process in an actual settling tank is reduced by turbulence and short-circuiting and increased by flocculation due to velocity gradients. Appropriate corrections must be made for these factors. An allowance must also be made for inlet and outlet losses and sludge storage. These influences have been discussed by Camp (1946), Dobbins (1944), Bloodgood (1956) and Ingersoll *et al.* (1956). The net effect of these factors results in a decrease of the overflow rate and an increase in the detention time over that derived from the laboratory analysis. As a general rule the overflow rate will be decreased by a factor of 1·25–1·75 and the detention period increased by a factor of 1·50–2·00.

SOLID–LIQUID SEPARATION   163

Average results of suspended solids and biochemical oxygen demand removals of primary treatment of domestic sewage are shown in Fig. 5-7. These data were obtained from Thomas and Dallas (1952), Fair and Geyer (1958), Phelps (1948) and Dorr Oliver Bulletin No. 6192 (1952). Industrial wastes may be expected

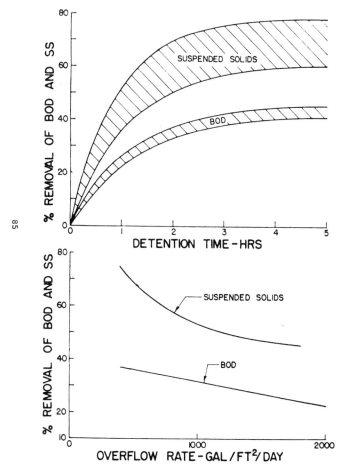

FIG. 5-7. Suspended solids removal from domestic sewage (from Fair and Geyer, 1958; Phelps, 1948; Dorr-Oliver Bull. No. 6192, 1952. Thomas and Dallas, 1952).

to vary markedly in both solids and BOD removal. One example taken from a paper and pulp waste is shown in Fig. 5-8.

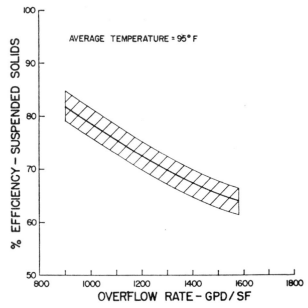

FIG. 5-8. Suspended solids removal characteristics for a pulp and paper mill waste.

Example 5-2. A laboratory settling analysis gave the following results:

Time, min	% Suspended solids removed at indicated depth		
	2 ft	4 ft	6 ft
10	47	27	16
20	50	34	43
30	62	48	47
45	71	52	46
60	76	65	48

Note:
$$C_0 = 430 \text{ p.p.m.}, \quad T = 29°C.$$

(a) Design a settling tank to remove 70 per cent of the suspended solids for 1 Mgal/day flow. (Apply appropriate factors); neglect initial solids effects.

(b) What removal will be attained if the flow is increased to 2 Mgal/day?

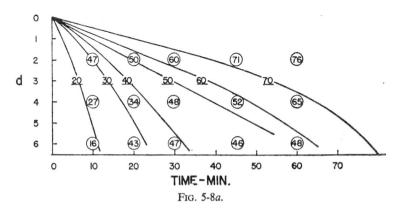

FIG. 5-8a.

The following data is developed in accordance with the procedure outlined above:

Time, min	Vel. ft/hr	Removal of SS	O.R. gal/ft²/day
12	30	34%	5400
23·2	15·5	49%	2790
32·5	11·1	55·8%	2000
62·0	5·8	67·1%	1040

These data result in the following plots:

For a flow of 1 Mgal/day the overflow rate for 70 per cent removal of suspended solids from above is 1000 gal/day/ft². Applying a scale-up factor of 1·5 the operating overflow rate is:

$$\text{O.R.} = 1000/1·5 = 666·7 \text{ gal/day/ft}^2$$

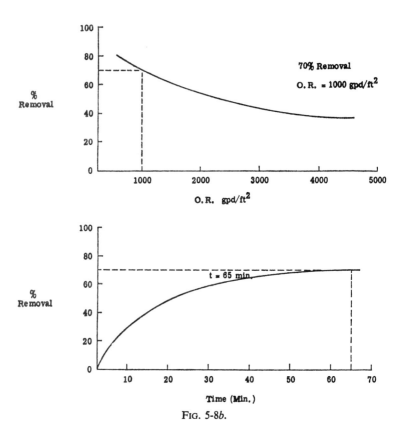

Fig. 5-8b.

The detention time is 65 min and applying a scale-up factor of 1·75

$$\text{Time} = 65 \text{ min} \times 1\cdot75 = 114 \text{ min}$$

$$\text{Surface area} = \frac{1 \text{ Mgal/day}}{666\cdot7 \text{ gal/day/ft}^2} = 1500 \text{ ft}^2$$

This is a 45 ft diameter tank.

$$\text{Effective depth} = \frac{1\cdot0 \times 114/60 \times 1/24 \times 1/7\cdot5}{1500} = 7\cdot05 \text{ ft}$$

For a flow of 2 Mgal/day:

O.R. = 2000 gal/day/ft² (theoretical)

Per cent suspended solids removal = 55

ZONE SETTLING AND COMPRESSION

In the settling of activated sludge, the floc adheres together and the mass of particles settles as a blanket forming a distinct interface between the floc and the supernatant. The settling process may be

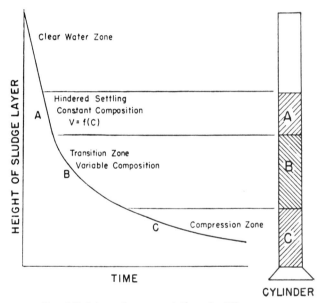

FIG. 5-9. Schematic representation of settling zones for activated sludge.

distinguished by three zones as shown in Fig. 5-9. During the initial settling period (A) the sludge floc settles at a uniform velocity under conditions of zone settling. The magnitude of this velocity is a function of the initial solids concentration. The concentration of solids during zone settling will remain constant until the settling interface approaches an interface of critical concentration. As the depth of settled sludge solids increases, the settling floc begins to press on layers below and a transition zone (B) occurs. Through the transition zone the settling velocity will decrease due to the increasing density and viscosity of the suspension surrounding the particles. A compression zone (C) occurs when the floc concentration becomes so great as to be mechanically supported by the layers of floc below. The solids concentration in the compression zone will be related to

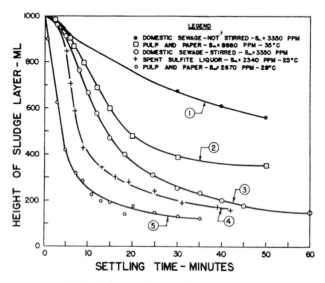

FIG. 5-10. Activated sludge settling characteristics.

the depth of sludge and the detention of the solids in this zone. Cummings *et al.* (1954) showed that for a given detention period a shallow compression zone depth will yield a higher underflow concentration than a deep depth. As the depth of the sludge blanket is increased, the sludge detention in the compression zone must also be increased for a specified underflow concentration. Fig. 5-10 shows settling curves for activated sludge under various conditions of concentration and mixing. Curves 1 and 3 illustrate the effect of stirring (4–5 rev/hr) on the same sludge. The unstirred sludge stratified and compacted at a much lower rate than the stirred sludge. Curves 2 and 5 show laboratory settling curves for stirred activated sludge from the oxidation of a pulp and paper mill waste at two concentrations and illustrate the effect of concentration on settling and compaction. It is significant to note that in many cases, particularly at higher concentrations, the sludge undergoes an initial period of flocculation as shown by Curves 3 and 4.

Zone settling is represented as the constant rate settling of the sludge–liquid interface which initially occurs in a batch sedimentation test. This settling rate is a function of the initial solids concentration and the flocculation characteristics of the suspension. As the initial concentration is increased the settling velocity of the sludge mass is decreased. The velocity-concentration curve for any sludge will depend on the physical and chemical properties of the sludge and on its flocculating characteristics. A particular sludge may alter its settling properties depending on the biological characteristics of the system (i.e. BOD loading, dissolved oxygen levels, etc). The zone settling velocity of a domestic sewage activated sludge over the range of mixed liquor suspended solids of 1000–4000 p.p.m. was found to be 6 to 20 ft/hr. Activated sludge from kraft mill oxidation and paper repulping waste oxidation showed a settling velocity range of 10 to 26 ft/hr over a concentration range of 3000–6000 p.p.m. and 1 to 8 ft/hr over a concentration range of 1000–2000 p.p.m. Camp (1946) has reported a range of zone settling velocity of 16·6 to 32·0 ft/hr for domestic sewage activated sludge.

When the sludge interface level in the hindered settling zone approaches the level of sludge in the compression zone a transition zone occurs. The transition zone is characterized by a critical concentration at which point compression of the settling sludge mass commences. The critical concentration has been found to vary widely with different sludges and to vary with initial solids concentration in most cases. The critical concentration can be estimated from a batch settling test by projecting tangents from the hindered settling zone and the compression zone and bisecting the angle produced by the two tangents. The point at which the bisected angle cuts the settling curve is the approximate compression point. The critical concentration at the sludge interface at the compression point is computed from the relationship (Talmadge and Fitch, 1955):

$$C_0 H_0 = C_2 H_1 \qquad (5\text{-}7)$$

where H_1 is the height the sludge would occupy if the entire sludge mass had the same concentration as the interface (C_2) at the compression point. H_1 is obtained by projecting a tangent at the compression point to the sludge height axis. The interfacial concentration at the compression point for a domestic sewage activated sludge was found to vary with initial solids concentration and showed a range of 4000–8000 p.p.m. over an initial solids concentration range of 1000–4000 p.p.m. (normal sewage plant operating range). These data are in substantial agreement with results reported by Camp (1946). Data on activated sludge from a kraft mill oxidation showed the compression point to be independent of initial solids concentration. The compression points from the non-stirred samples were slightly lower than those for the stirred samples.

During sludge compression, water is squeezed from the compacting sludge mass. The rate of compaction is low because the displaced fluid has to flow through the small void space within the sludge mass, and progressively decreases with time as the resistance to liquid flow increases. A relationship has been developed to approximate the rate of sludge thickening with time (Roberts, 1934);

$$-\frac{dH}{dt} = K(H - H_\infty) \qquad (5\text{-}8)$$

Equation (5-8) may be integrated:

$$2\cdot 3 \log \frac{(H_c - H_\infty)}{(H - H_\infty)} = K(t - t_c) \qquad (5\text{-}9)$$

In Equation (5-9) H_c is the height of the sludge layer at the compression point and H_∞ is the height the sludge will occupy after compression is complete.

Equation (5-9) may also be expressed in terms of sludge concentration:

$$2\cdot 3 \log \frac{(C_\infty - C_c)}{(C_\infty - C)} = K(t - t_c) \qquad (5\text{-}10)$$

C_∞ is determined from laboratory data on cylinder tests and plotting $\log (H - H_\infty)$ vs. time. The value of H_∞ is adjusted by trial and error until a straight line is attained on the plot. C_∞ is computed from the relationship:

$$C_0 H_0 = C_\infty H_\infty$$

Behn (1957) has shown that the rate coefficient K, in Equation (5-10) is inversely proportional to the depth of sludge in the compression zone. Therefore, if laboratory thickening data are to be employed for scale-up in thickener design, the ratio of times required in the compression zone in the laboratory and prototype is proportional to the ratio of the respective depths. Compression data of an activated sludge from a pulp and paper waste are shown in Fig. 5-11. As shown in Fig. 5-11, the compression characteristics are related to the sludge index.

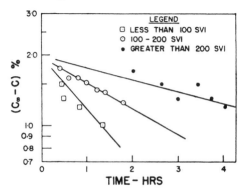

FIG. 5-11. Secondary sludge compaction characteristics.
SVI refers to the Sludge Volume Index.

Clarifier Design

The performance of secondary sedimentation units in biological processes is related to both the physical and chemical nature of the sludge and to the hydraulic characteristics of the sedimentation basin. Sludge characteristics have been discussed by Rudolfs and Lacy (1934) and Heukelekian (1956). Factors affecting hydraulic design have been discussed by Anderson (1945), Sawyer (1956) and Walker (1957). Theoretical discussions of flocculent sludge settling have been given by Coulson and Richardson (1956) and Fitch (1957).

A clarifier employed in biological treatment processes serves the dual functions of clarifying the liquid overflow and thickening the sludge underflow. The clarification capacity of the unit is related to the settling velocity of the sludge in which the velocity of the sludge interface must be greater than the vertical rise velocity of the liquid at any level. This settling rate may be approximated by the change in interface level in batch sedimentation tests with time. The thickening capacity is related to the depth of sludge in the basin and the

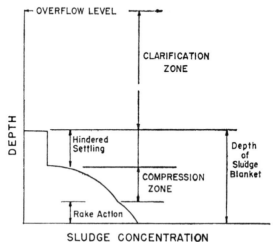

FIG. 5-12. Clarifier settling zones.

time the sludge is in the compression zone. Thickening is enhanced by the hydraulic movement of the sludge blanket and the action of the rakes which break up the arching of the settled sludge. In clarifier operation, where the sludge is introduced through a feed well there may be substantially no free settling zone. If the clarifier becomes overloaded a zone of constant composition will develop in the basin and will eventually rise and overflow the basin. These settling zones are shown in Fig. 5-12.

The size of secondary clarifiers in bio-oxidation systems is therefore related to three design factors, namely (a) the area required for clarification over the operating mixed liquor suspended solids range. This in turn is related to the allowable overflow rate such that the vertical liquid rise rate at any level is less than the solids subsidence rate at that level. (b) The area and volume requirements to produce

SOLID–LIQUID SEPARATION   173

by thickening an underflow of a desired concentration. (c) The permissible retention of the settled sludge in the basin as dictated by its biological properties. In any given case the tank area and volume will be controlled by one of these three factors.

The overflow rate required for clarification is computed from the free settling velocity of the sludge. As has been previously discussed the rate of sludge subsidence must be greater than the rate of liquid rise. This rate is computed using only the overflow from the basin. This overflow rate may be approximated over a wide concentration range (usually 1000–5000 p.p.m.) using laboratory data.

The area requirements for sludge thickening may be computed by the methods developed by Coe and Clevenger (1913) and Talmadge and Fitch (1955). Considering the concentration range over which the basin will function the required unit area may be computed from the relation:

$$\text{U.A.} = \frac{(1/C_2) - (1/C_u)}{V_2} \tag{5-11}$$

In Equation (5-11) the units of C_2 and C_u are lb/ft^3 and the units of V_2 is ft/day.

An area should be computed for all concentration ranges and the largest area used for design. Usually the area determined at the critical concentration (compression point) will be the controlling area.

The method of Talmadge and Fitch (1955) involves a settling test in which the sludge height at various times is plotted against the time to attain that height. A tangent is drawn to the point of compression. The intersection of this tangent with the line representing the desired underflow concentration C_u determines the time t_u. This is the time required to attain the desired underflow concentration. A wide range of initial solids should be evaluated and the maximum area in the expected operating range selected.

The required area may be computed from the relationship:

$$\text{U.A.} = \frac{t_u}{C_0 H_0} \tag{5-12}$$

When considering biological sludges the underflow concentration which can be developed by compression may be limited. The sludge in a clarifier will consume the available oxygen and progress into an anaerobic state. In sludge with a high rate of activity this may result in gasification and floating of the sludge to the surface of the unit and the allowable detention of the sludge in the clarifier will be very short. For example, a sludge from a pharmaceutical waste oxidation (BOD = 5000 p.p.m.) gasified and floated to the surface after 15–30 min sedimentation after 3 hr aeration. When the aeration period was extended to 12 hr flotation did not occur even after 4 hr sedimentation. A limiting retention period of the sludge in the clarifier will be established which will in turn produce a limiting underflow concentration. This limiting underflow concentration has been found to vary from 0·5 per cent for a highly active sludge from a cannery waste oxidation to 3·0 per cent for pulp and paper waste oxidation. A method employing redox potential has been developed by Nussberger (1955) to determine the retention period of activated sludge in final clarifiers. Since unit areas and settling velocities are determined from laboratory experiments, scale-up factors are required to convert these criteria to plant scale.

Example 5-3. The following data was obtained for secondary sludge settling.

(a) Free settling data:

V_s ft/hr	C_0 p.p.m.
24·3	1000
12·0	2000
7·0	3000
4·8	4000
3·8	5000

(b) Sludge compaction data (settling tank):

T, hr	C_u
0·5	0·5
1·0	0·65
2·0	0·95
3·0	1·15

SOLID–LIQUID SEPARATION   175

(c) A typical laboratory settling study gave the following data:

$$C_0 = 2600 \text{ p.p.m.}; \quad H_0 = 1\cdot15 \text{ ft}$$

T, min	Height of sludge interface (mi)
0	1000
1	900
3	620
5	420
7	320
11	220
15	195
26	150
30	130

A. Using this data, compute the overflow rate of $C_0 = 2600$ p.p.m. and $C_u = 10,000$ p.p.m. (flow-through basis).

B. Check this overflow rate for a C_0 of 4000 p.p.m. (free settling).

C. Compute the sludge blanket depth.

D. Specify the controlling overflow rate for this case.

A—For the laboratory settling data:

$$C_0 H_0 = C_u H_u$$
$$C_u = 10,000 \text{ p.p.m.}$$
$$H_u = 260 \text{ ml}$$
$$t_u = 10 \text{ min}$$

$$U.A. = \frac{t_u}{C_0 H_0}$$

$$= 10 \text{ min} \times \frac{1 \text{ day}}{1440 \text{ min}} \times \frac{1}{1\cdot15 \text{ ft}} \times \frac{1}{2600 \text{ p.p.m.}}$$

$$\times \frac{1 \text{ ft}^2}{62\cdot4 \text{ lb}}$$

(assuming mixed liquor solids has same s.g. as water)

$$= 0\cdot037 \text{ ft}^2/\text{lb per day}$$

The unit area may be converted to overflow rate by the following relations:

$$\text{OR} = \frac{120}{(\text{U.A.})(C_0 \text{ gr/l.})} \cdot \frac{C_u - C_0}{C_u}$$

(flow through) $\text{OR} = \dfrac{120}{(0\cdot037)(2\cdot6)} \dfrac{7\cdot4}{10} = 930 \text{ gal/day/ft}^2$

B—For hindered settling

$C_0 = 4000$ p.p.m.
$V_s = 4\cdot8$ ft/hr $\times$ 180 $= 865$ gal/day/ft²

C—Sludge blanket depth, for 10,000 p.p.m.

$t = 2\cdot25$ hr

$$D = \frac{T}{\rho C_u \text{ U.A. } 24} = \frac{2\cdot25 \text{ hr}}{(24)(62\cdot4)(0\cdot037)(0\cdot01)} = 4\cdot10 \text{ ft}$$

Hydraulic Considerations

Density currents are induced in final settling tanks due to the fact that the specific gravity of the sludge mixture is greater than the clarified water in the tank. The activated sludge, on entering a settling tank, falls almost vertically and then flows along the bottom of the tank toward the outlet end or wall. Currents in the bottom of the tank will establish secondary currents in the water layers above in a reverse direction. This phenomena may lead to a carry-over of sludge in the peripheral weirs from up-welling currents. This effect is minimized by increasing weir lengths to decrease lineal overflow rate and by proper inlet design.

Inlet and Outlet Devices

The purpose of an inlet device is to distribute the flow uniformly across the width and depth of the settling tank. The outlet device is likewise designed to collect the effluent uniformly at the outlet end of the tank. Inlets and outlets of good design reduce the short-circuiting characteristics of a tank. A uniform distribution of inflow is effected by making the head loss through the inlet ports large by comparison with the friction loss in the inlet channel in which the velocity should be maintained above 0·5 ft/sec. Thus a small difference in head from one port to another will have little effect on the relative discharge of the ports. If the head loss through the ports is equal to or greater than 5·3 times the friction in the channel, then the discharge through the near port will vary not more than 10 per cent from that through the far port. The hydraulic design of these units should be based on maximum flow and should be checked for average and minimum conditions.

Outlet devices are designed to collect the effluent uniformly at the outlet with minimal take-off velocities to prevent carry-over of sludge to the effluent channels. Increased weir length in rectangular basins is usually required, by extending the effluent channels back into the basin or by providing multiple effluent channels. Circular basins usually have sufficient periphery to insure low take-off velocities, although in-board weirs are sometimes advisable. Weir rates of the order of 10,000 to 15,000 gal/day/ft of weir appear to be satisfactory for secondary settling of activated sludge, weir rates up to 50,000 gal/day/ft may be employed for primary settling.

Typical inlet and outlet devices are shown in Fig. 5-13a.

Clarifier Mechanisms

In most rectangular clarifiers scraper flights extending the width of the tank move the settled sludge toward the inlet end of the tank at a speed of about one foot per minute. Some designs move the sludge toward the effluent end of the tank corresponding to the direction of flow of the density current.

Most circular tanks are designed for center sludge withdrawal with a bottom floor slope of one inch per foot. The flow of sludge to the center well is largely hydraulically motivated by the collection mechanism serving to overcome inertia and avoid sludge adherence to the tank bottom.

The types of clarifier mechanisms commonly employed include the plow type, the rotary hoe type, and the vacuum drawoff type. The plow type mechanism employs staggered plows attached to two opposing arms which move about 10 ft/min. The rotary hoe mechanism consists of a series of short scrapers suspended from a rotating supporting bridge on endless chains which make contact with the tank bottom at the periphery and move to the center of the tank. The vacuum type mechanism consists of a series of orifices attached to rotating arms which continuously withdraw sludge from all

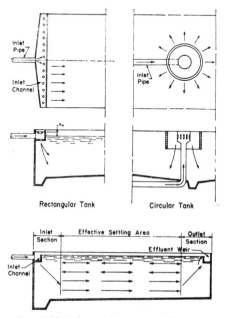

FIG. 5-13 (a–b). Inlet and outlet devices.

sections of the tank floor. Various types of collecting mechanisms are shown in Fig. 5-14.

Thickening

Gravity thickeners are employed to thicken primary and secondary sludges prior to final disposal by drying or vacuum filtration. Thickening is accomplished by compression and compaction. During compression liquid flows through a network of capillaries to the surface of the slurry. During the final stages of compaction the settling rate is governed by diffusion. Roberts (1934) has shown

SOLID-LIQUID SEPARATION 179

that the thickening process can be estimated from Equation 5-10. The logarithmic plot of Equation 5-10 will have a discontinuity at the beginning of the compression zone. Most gravity thickeners have picket fence type of rake mechanism to break up the arching action of the settled sludge.

Thickeners are frequently designed in terms of unit area. Schroepfer (1959) showed that the unit area required to thicken sludge from the anaerobic treatment of packing house waste with an SVI of 50–100 was 50–85 lb/day per ft^2 of thickener area. Sludge from the anaerobic treatment of dairy waste with an SVI of 100–300 was thickened at a loading of 25–35 lb/day per ft^2. Nelson and Budd (1959) have recommended thickener loadings of 20 lb/ft^2 per day to produce a 12 per cent solids underflow concentration of 8 per cent solids for trickling filter sludge and 8 lb solid ft^2/day to produce a sludge concentration of 6 per cent solids from activated sludge. Experimental studies at Greenwich, Conn. produced an underflow of 8·3–10 per cent solids at thickener loadings of 17·9–23·0 lb solids/ft^2/day. Data from Riverside, Cal. produced a thickened sludge of 7·4–7·9 per cent solids at a loading of 6·6–6·8 lb solids/ft^2 per day.

FLOTATION

Flotation is employed for the removal of suspended solids from wastes and for the separation and concentration of biological flocculent sludges. In current flotation practice, the waste flow or a portion of clarified effluent is pressurized to 40–60 p.s.i. in the presence of sufficient air to approach saturation. Minute air bubbles are released from solution when this air-saturated liquid is mixed with a sludge mixture at atmospheric pressure. The sludge flocs and suspended solids are floated by these minute air bubbles which attach themselves to and are enmeshed in the floc particles. The air–sludge mixture rises to the surface where it is skimmed off. The clarified liquor is removed from the bottom of the flotation unit at which time a portion of the effluent may be recycled back to the pressure chamber. Superior effluent quality and economy in power will frequently accrue from pressurizing clarified effluent rather than the waste flow. This is particularly true for flocculent sludges which are dispersed by high shearing stresses. Activated

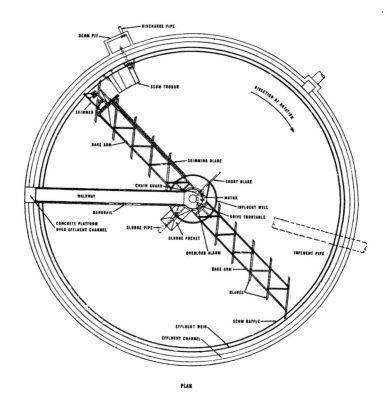
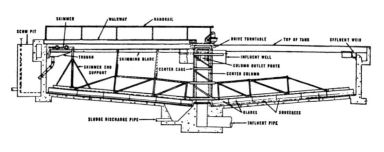

FIG. 5-14. (a). Plan and elevation—circular clarifier. (*Courtesy of Dorr-Oliver, Inc.*).

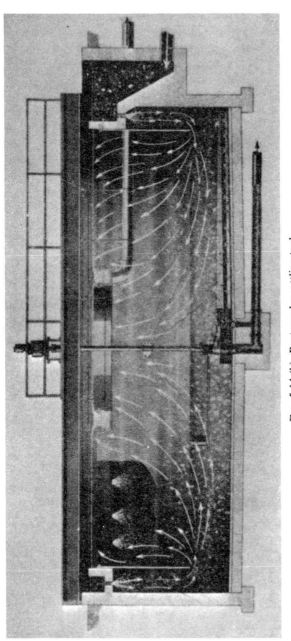

FIG. 5-14 (b). Rectangular settling tank. (*Courtesy of Chain Belt Co.*).

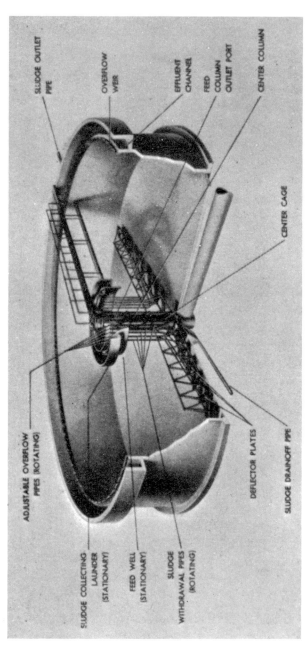

Fig. 5-14 (c). Circular clarifier mechanism. (*Courtesy of Dorr-Oliver, Inc.*).

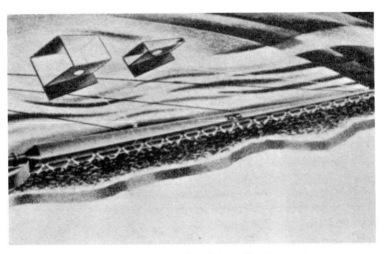

FIG. 5-14 (d). Suction sludge collection. (*Courtesy of Chain Belt Co.*)

sludge is a friable flocculent sludge. When exposed to high shearing forces a portion of the floc is dispersed into fine particles which are difficult to separate. Cases in which mixed liquor was pressurized indicated progressive deterioration of effluent quality with increasing volumes of pressurized sludge.

Air Solubility and Release

The saturation of air in water is directly proportional to the pressure and inversely proportional to the temperature. Pray *et al.* (1952) and Frohlich (1931) found that oxygen and nitrogen solubility in water follows Henry's law over a wide pressure range. Recent experiments on water and several industrial wastes showed that, while a linear relation existed between pressure and solubility, the slope of the pressure–solubility curve varied, depending on the nature of the constituents present. Calculated and observed solubility relationships are shown in Fig. 5-15 (Vrablik, 1959). The quantity of air which will be released from solution when the pressure is reduced to one atmosphere under turbulent conditions is a function of the temperature, pressure and volume of pressurized flow. It is usually expressed as standard cubic feet of air per gallon or pounds of air per gallon released from an air saturation solution. Fig. 5-15 shows the pressure, temperature, air release relationship in pure water and in other solutions.

Rise Rates

Studies in laboratory flotation cells have shown that the mass velocity of rise of the air bubbles released decreases with increasing pressure. Vrablik (1959) has shown that bubbles released after pressurization to 20–50 lb/in^2 range in size from 30–120 μ. The velocity of rise of the particles closely follows Stokes' law. Rise rates during the free separation phase have been observed over a range of one inch per minute to five inches per minute in the flotation cell. In the flotation of biological sludge the rate of sludge blanket rise will increase with an increasing air-to-solids ratio due to an increasing number of bubbles enmeshed in the sludge blanket. Hurwitz and Katz (1959) in the flotation of activated sludge of 0·91 per cent solids at 40 p.s.i. pressure observed a free rise rate of 0·3 ft/min, 1·2 ft/min and 1·8 ft/min for recycle ratios of 100 per cent, 200 per cent and 300 per cent respectively. Results from the laboratory flotation cell indicated that initial vertical rise rate may vary from 2 in/min to 5 in/min for domestic sewage activated

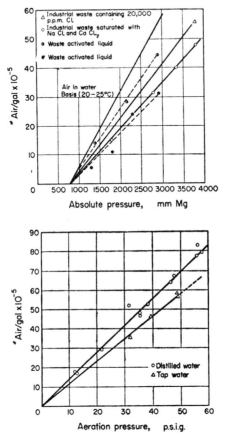

Fig. 5-15. Air solubility and release for various pressures. (after Vrablik, 1960).

sludge over an air-to-solids ratio range of 0·015 to 0·06 lb air/lb solids. The initial rise rate was observed to vary with the character of the sludge being separated. Activated sludge from pulp and paper waste oxidation had an initial rise rate of 8 in/min to 10 in/min over an air-to-solids ratio range of 0·15 to 0·25 lb of solids. Vrablik (1959) showed that the unit area methods developed for thickener design could be used to obtain relationships between float solids concentration and volume of float.

Flotation Design

The primary variables for flotation design are pressure, recycle ratio, feed solids concentration and retention period. The quantity of air release from solution per unit volume of pressurized liquid is directly proportional to the pressure. When other process variables are maintained constantly increasing the pressure will result in decreasing effluent suspended solids and increasing solids concentration in the floated sludge.

The effluent-suspended solids decrease and the concentration of sludge in the float increases with increasing detention period. Where the flotation process is employed primarily for clarification, a detention period of 20–30 min is adequate for separation and concentration. Rise rates of 1·5–4·0 gal/min/ft^2 are commonly employed. When the process is employed for thickening, longer retention periods are necessary (120 min) to permit the sludge to compact. Hurwitz and Katz (1959) obtained maximum concentrations of 5·6 per cent, 7·7 per cent and 10·8 per cent for activated sludge, a mixture of 50 per cent activated and 50 per cent primary sludge and primary sludge respectively.

The principle components in a flotation system are: a pressurizing pump, air injection facilities, a retention tank, a back-pressure regulating device and the flotation unit. These components are shown in Fig. 5-16. The pressurizing pump creates an elevated pressure to increase the solubility of air. Air injection is usually accomplished through an injector on the suction side of the pump.

The air and liquid are mixed under pressure in a retention tank. The detention in the pressurizing tank is usually 1–2 min. The per cent of saturation obtained is related to both the time and the air volume supplied to the retention tank. Experimental investigation (Hays, 1956) has shown air saturation to vary from 20–35 per cent to 36–98 per cent at 0·6–1·4 ft^3 air/100 gal flow at 1·0 and 2·5 min retention respectively.

A back-pressure regulating device maintains a constant head on the pressurizing pump. Various types of special valves are employed for this purpose. The flotation unit may be either circular or rectangular with a skimming device to remove the thickened, floated sludge. A flotation unit is shown in Fig. 5-17.

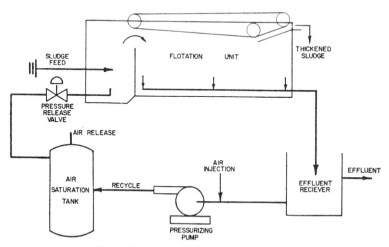

Fig. 5-16. Dissolved air flotation system.

In the evaluation of flotation variables for process design, it is convenient to employ a dimensionless air-to-solids ratio. This ratio is defined as the lb of air released from the pressurized recycle divided by the lb of suspended solids treated. The A/S can be computed from the following equation, depending upon the pressure, solubility and volume of recycle:

$$\frac{A}{S} = \frac{Cs_a (f'P - 1) R}{S_a Q} \qquad (5\text{-}13)$$

$C =$ Constant
$s_a =$ Solubility of air in water at standard conditions
$P =$ Pressure, atmospheres
$f' =$ Percent solubility
$R =$ Pressurized recycle rate
$S_a =$ Suspended solids content
$Q =$ Waste flow

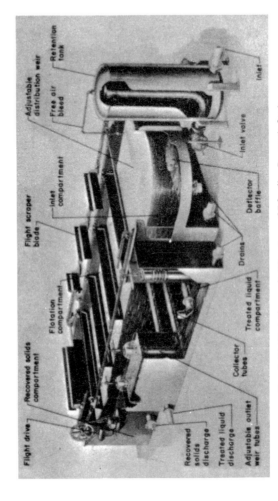

FIG. 5-17. Flotation unit (*Courtesy of Pfaudler-Permutit Inc.*).

Example 5-4. A pilot plant flotation operation indicated the optimum air/solids ratio to be 0·04 lb air/lb solids. If a waste to be treated has 250 p.p.m. suspended solids content compute the per cent recycle to be pressurized to 60 p.s.i. at 20°C. An air rate of 1·4 ft³/100 gal will be used with a retention of 2·5 min.

$$\frac{A}{S} = \frac{Cs_a (f'P - 1) R}{S_a Q}$$

From the data of Hays (1956) $f = 0.68$ (for other operating conditions than those described by Hays f would have to be measured).
and

$$R \text{ (mg)} = \frac{0 \cdot 834 \, A/s S_a Q}{s_a (f'P - 1)}$$

in which

$$P = \frac{60 + 14 \cdot 7}{14 \cdot 7} = 5 \cdot 1 \text{ atm}$$

$$s_a = 18 \cdot 7 \text{ cm}^3/1 \text{ air solubility}$$

$$A/S = 0 \cdot 04$$

$$R = \frac{(0 \cdot 834)(0 \cdot 04)(250)(1)}{(18 \cdot 7)(0 \cdot 68 \cdot 5 \cdot 1 - 1)}$$

$$= 0 \cdot 160 \text{ mg}$$

Scale-up factors would have to be employed to this computed value.

The effluent suspended solids and the sludge float concentration were related to the air-to-solids ratio for a low index domestic sewerage activated sludge, a high index domestic sewage activated sludge and activated sludge from pulp and paper and chemical waste oxidation as shown in Figs. 5-18 and 5-19. These data were obtained over a range of pressures from 30 to 80 lb/in² and a range of recycle of 25 to 150 per cent.

The relationships obtained vary with the source and character of sludge as might be expected. Increasing the air–solids ratio beyond an optimum value will result in no substantial reduction in effluent suspended solids or increase in sludge concentration. For data obtained on several activated sludges the optimum air–solids ratio varied from 0·03 to 0·1.

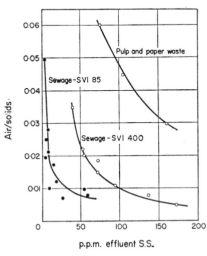

FIG. 5-18. Variation in effluent suspended solids with air to solids ratio.

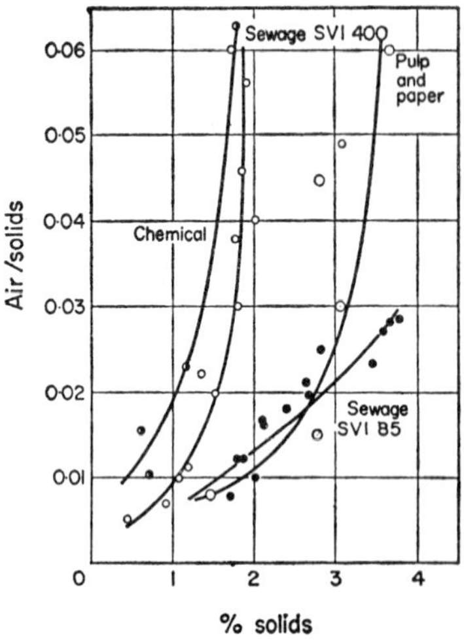

FIG. 5-19. Variation in float solids concentration with air to solids ratio.

REFERENCES

1. ANDERSON, N. E., *Sew. Works J.* **17**, 50 (1945).
2. BEHN, V. C., *Proc. Amer. Soc. Civil Engrs.*, **83** No. S.A. 5 (1957).
3. BLOODGOOD, D. E., BOEGLY, W. J. and SMITH, C. E., *Proc. Amer. Soc. Civil Engrs.*, **82**, 1083 (1956).
4. CAMP, T. R., *Trans. Amer. Soc. Civil Engrs.*, **11**, 895 (1946).
5. CAMP, T. R., *Sew. & Ind. Wastes*, **25**, 1 (1953).
6. COE, H. S. and CLEVENGER, G. H., *Trans. Amer. Inst. Mining Met. Engrs.*, **55**, 356 (1916).
7. COULSON, J. M. and RICHARDSON, J. F., *Chem. Enging*, p. 511, McGraw Hill Book Co., New York (1955).
8. CUMMINGS, E. W., PRIESS, C. E. and DeBORD, C., *Ind. Eng. Chem.*, **46**, 8, 1164 (1954).
9. DOBBINS, W. E., *Trans. Amer. Soc. Civil Engrs.*, **109**, 629 (1944).
10. DORR-OLIVER, Inc., Bull No. 6192 (1952).

11. FAIR, G. M. and GEYER, J. C., *Water Supply and Waste Water Disposal*, New York, John Wiley and Sons, Inc. (1954).
12. FROHLICH, R., *Ind. Eng. Chem.*, **23**, 548 (1931).
13. FITCH, E. B., *Biological Treatment of Sewage and Industrial Wastes*, Vol. II, (Ed. by MCCABE, B. J. and ECKENFELDER, W. W.) Reinhold Pub. Corp. (1958).
14. HAYS, T. T., *Sew. & Ind. Wastes*, **28**, 1, 100 (1956).
15. HAZEN, A., *Trans. Amer. Soc. Civil Engrs.*, **53**, 45 (1904).
16. HEUKELEKIAN, H., *Sew. & Ind. Wastes*, **28**, 1, 100 (1956).
17. HURWITZ, E. and KATZ, W. J., "Laboratory Experiments on Dewatering Sewage Sludges by Dissolved Air Flotation" (Unp.) (1959).
18. INGERSOLL, A. C., MCKEE, J. E. and BROOKS, N. H., *Trans. Amer. Soc. Civil Engrs.*, **82**, 1083 (1956).
19. KALINSKE, A. A., *J. Amer. Water Works Assoc.*, **40**, 113 (1946).
20. NELSON, F. G. and BUDD, W. I., *Proc. Amer. Soc. Civil Engrs, SED* (Nov. 1959).
21. NUSSBERGER, F. E., *Sew. and Ind. Wastes*, **25**, 12, 1003 (1953).
22. O'CONNOR, D. J. and ECKENFELDER, W. W., *Biological Treatment of Sewage and Industrial Wastes*, Vol. II (Ed. by MCCABE, B. J. and ECKENFELDER, W. W.) Reinhold Pub. Corp. (1958).
23. PHELPS, E. P., *Public Health Engineering*, New York, John Wiley and Sons, Inc. (1948).
24. PRAY, H. A., *Ind. Eng. Chem.*, **44**, 1146 (May, 1952).
25. ROBERTS, E. J., *Mining Eng.*, **1**, 61 (1946).
26. RUDOLFS, W. and LACY, J., *Sew. Wks. Jour.*, **6**, 4, p. 647 (1934).
27. SAWYER, C. N., *Biological Treatment of Sewage and Industrial Wastes*, Vol. II (Ed. by MCCABE, B. J. and ECKENFELDER, W. W.) Reinhold Pub. Corp., New York (1956).
28. SHROEPFER, G. J. and ZIENKE, N. R., *Sew. and Ind. Wastes*, **31**, 12, 697 (1959).
29. TALMADGE, W. P. and FITCH, E. B., *Ind. Eng. Chem.*, **47**, 38 (1955).
30. THOMAS, H. A., Jr., and DALLAS, J. L., *J. Boston Soc. Civil Engrs.*, **39**, 354 (1952).
31. VRABLIK, E. R., *Proc. 14th Ind. Waste Conf. Purdue Univ.* (1959).
32. WALKER, D. J. and DREIER, D. E., *Biological Treatment of Sewage and Industrial Wastes*, Vol. II (Ed. by MCCABE, B. J. and ECKENFELDER, W. W.) Reinhold Pub. Corp. (1958).

CHAPTER 6

AEROBIC BIOLOGICAL TREATMENT PROCESSES

BIOLOGICAL waste treatment facilities cover a wide spectrum of process applications from the aerated lagoon and stabilization basin with long retention periods and low solids to the high rate activated sludge process with short aeration periods and high aeration solids. The selection of method depends primarily on the characteristics of the wastes, geographical location and effluent requirements of the regulatory agency. The various processes in use today are discussed in this chapter along with design procedures for specific applications.

LAGOONS AND STABILIZATION BASINS

Lagoons and stabilization basins may be classified into 3 general types:

Type I—Large holding reservoirs with detention periods in the order of months. Oxygen is supplied by means of surface aeration and by dilution of the waste with non-contaminated water.

BOD removal is by bio-flocculation, oxidation and reduction. Active anaerobic digestion of bottom sludge deposits occurs.

Type II—Oxidation ponds with detention periods ranging from less than a week to 6 weeks.

This type of pond is sub-divided into two classifications, the facultative pond and the high rate pond. In the facultative pond, BOD is removed by sedimentation, bioflocculation and aerobic oxidation. Oxygen is obtained from algal growths. Surface aeration is a secondary source of oxygen. Settled sludge deposits undergo decomposition by acid and methane fermentation. A balance between the aerobic and anaerobic processes is essential to minimize odors. Ponds are usually 2–5 ft in depth and may be constructed with no overflow (evaporation and seepage exceeds in-flow) or with a regulated overflow.

The high rate pond is constructed with shallow depths (½–1 ft) to permit light to penetrate to the bottom. Fully aerobic conditions are maintained by mixing. In this system, the organic matter is oxidized by bacteria, which are utilizing oxygen, released by the algal growths. The algae employ solar energy to synthesize cell substance from carbon dioxide and ammonia.

Type III—Aerated lagoons with detention periods ranging from a few days to 2 weeks, depending upon the efficiency desired. Oxygen is supplied by diffused or mechanical aeration systems, which also cause sufficient mixing to induce a significant amount of surface aeration. Depths from 6 to 15 ft are common. Organic matter is oxidized by bacterial action.

Oxidation Ponds

In oxidation ponds, the rate of oxidation of organic matter by bacteria exceeds the rate of natural surface reaeration and oxygen must be supplied by algae growths. The algae employ solar energy for cell synthesis, using CO_2 as a carbon source and ammonia as a nitrogen source, and release molecular oxygen to solution. Since light penetration is essential for algal growth, the depth of these basins is limited.

Depending on the depth and geometry of the basin, the process may be fully aerobic or be a combination of aerobic and anaerobic processes.

In domestic sewage treatment, BOD is primarily removed by sedimentation and bioflocculation. This clarification requires a rich and varied population of microorganisms and is aided by the presence of invertebrates such as rotifers, etc.

Oswald (1960) has shown that up to 85 per cent of the suspended and dissolved organics are deposited on the basin bottom in 4 hr. Clarification is also aided by auto flocculation at high pH values. In this case inorganic salts precipitate enmeshing algae and bacteria which subsequently settle to the bottom of the basin.

Aerobic oxidation in ponds was found by Oswald *et al.* (1958) to follow the relationship:

$$C_{11}H_{29}O_7N + 14O_2 + H^+ \rightarrow 11CO_2 + 13H_2O + NH_4^+.$$

The weight of oxygen required to oxidize organic matter would therefore be 1·56 times the weight of organic matter oxidized.

Oxygen is supplied to the pond by atmospheric reaeration and through photosynthesis. The quantity of oxygen by reaeration was approximated by Imhoff and Fair (1940) to be:

$$R = 0\cdot 0271 \,.\, (a) \,.\, (d) \,.\, (D_0) \tag{6-1}$$

in which R is the reaeration in lb O_2 per acre/day, d is the basin depth in feet, and D_0 is the mean daily saturation deficit. The factor a was assigned an arbitrary value of 20 by Oswald, 1960. At a mean deficit of 6 p.p.m., 6 lb of O_2 per acre/day would be absorbed in a pond of 3 ft mean depth. The reaeration is increased by wind or wave action.

Oxygen obtained by photosynthesis by algae was found to follow the relationship (Oswald, 1960)

$$NH_4^+ + 7\cdot 6CO_2 + 17H_2O = C_7H_{8\cdot 1}O_{2\cdot 5}N + 7\cdot 6O_2 + 15\cdot 2H_2O + H^+ - 886 \text{ Kcal/g}$$
(algae)

Approximately 3·68 cal are fixed for each mg of oxygen liberated and about 1·67 mg of oxygen is liberated for each mg of algae synthesized. The energy source is the sun.

The types of algae found active in oxidation ponds are—*Chlorella, Scenedesmus, Euglena* and *Chlamydomonos.*

Oswald (1960) showed that 80–90 per cent of the soluble organics end up in a sludge which deposits on the pond bottom. This is equivalent to 1·25 lb VSS per lb BOD removed. This sludge undergoes decomposition and methane fermentation in anaerobic and facultative ponds and aerobic oxidation in high rate ponds.

Oxidation ponds may be designed on the basis of any one of three criteria:

(1) Population per acre

(2) Hydraulic loading, in/day

(3) Organic loading, lb BOD/acre per day

Anaerobic Ponds

Anaerobic ponds are usually constructed with a depth of 8–12 ft in order to reduce heat losses. Oswald (1960) has suggested a length to width ratio of 4 : 1 in order to deposit more sludge at the influent end where temperature conditions are more favorable for anaerobic decomposition.

Facultative Ponds

In facultative ponds, basin depth may vary from 2–5 ft. In general the maximum concentration of BOD which can be treated is 300 p.p.m. to avoid excessively shallow depths. BOD's more concentrated than this require dilution. In some cases, recirculation is desirable to mix the influent waste with the oxygen and the algal growths.

The allowable lagoon loading will depend upon the rate of BOD reduction, the algal activity and the treatment efficiency desired. Herman and Gloyna (1958) developed a design formulation for sewage treatment employing a 2–3·5 ft depth.

$$V = (5 \cdot 37 \times 10^{-8} Nq \; Y \; 1 \cdot 072^{35-T}) \qquad (6\text{-}2)$$

in which V = lagoon volume in acre—ft
Nq = sewage flow in gal/day
Y = influent BOD in p.p.m.
T = temperature in °C

The performance of facultative oxidation ponds have been summarized by Herman and Gloyna (1958). For field installations in Texas with a depth of 2 ft they developed the relationship

$$\text{per cent BOD removal} = \frac{100}{(1 - 0 \cdot 04 \, (L)^{0 \cdot 57})} \qquad (6\text{-}3)$$

in which L is expressed in lb BOD/acre (day). Data from controlled laboratory experiments was correlated according to the relationship:

$$\text{per cent BOD removal} = 100 - 0 \cdot 05 L \qquad (6\text{-}4)$$

High Rate Ponds

Oswald and Gotaas (1957) have developed a design formulation relating the oxygen production by an algal growth to pertinent process variables:

$$D = \frac{h\,C_c\,d}{F\,1000\,S} \qquad (6\text{-}5)$$

in which D = detention period in days

h = unit heat of combustion of the algae, cal/mg

d = lagoon depth

F = efficiency of solar energy conversion

S = solar insolation expressed in Langley's (gm-cal)/(cm^2) (day)

C_c = algae concentration in mg/l.

The depth of an oxidation pond depends upon the concentration of the algal suspension and on the physical characteristics of the waste with respect to light transmission. The waste mixture has been found to absorb light in accordance with the Beer–Lambert law. If all the available light is absorbed in the basin such that the light intensity is just zero at the bottom of the basin it was shown that:

$$d = \frac{\log_e I_i}{C_c\,a} \qquad (6\text{-}6)$$

in which I_i = incident light intensity

C_c = algae concentration

a = specific absorption coefficient

d = depth through which light penetrates.

In most cases a may be expected to vary from 1×10^{-3} to 2×10^{-3} and will depend on the algal species and pigmentation and C_c from 100 to 300 p.p.m. The incident light intensity, I_i, will usually vary from 200–10,000 ft candles. The concentration of algae C_c, can be estimated from the relationship:

$$C_c = \frac{L_t\,O_t}{1\cdot 67} \qquad (6\text{-}7)$$

in which L_t is the ultimate BOD satisfied and O_t the fraction of oxygen produced which is used. (O_t varies from 1·2–1·6.)

The amount of solar energy, S, which reaches the earth's surface is related to geographical, meteorological and astronomical phenomena and is subject to wide seasonal and daily variations. It has been possible, however, to predict average monthly values for various geographical locations. These have been summarized by Oswald and Gotaas (1957). The nitrogen requirement for algal growth can be computed from the relationship:

$$C_c = 10 \times N \qquad (6\text{-}8)$$

in which C_c = maximum algal cell concentration
N = nitrogen required

Only a portion of the light is converted to fixed energy in the form of algal cells.

Equation (6-8) is based on 80 per cent nitrogen recovery from the waste and an 8 per cent nitrogen content in the algae. Algae are reported to contain 1·5 per cent phosphorus (dry weight). A concentration of 6 p.p.m. phosphorus in the waste will sustain 400 p.p.m. of algae. Magnesium and potassium are essential elements for algal propagation. Algal cells will contain 0·5 per cent K and 1 per cent Mg.

The light conversion efficiency, F, is a function of light, time, nutrient and temperature. Oswald (1960) has shown the conversion efficiency to be related to the detention period, the light efficiency, the diurnal illumination, the BOD applied and temperature. Tables have been prepared by Oswald (1960) to permit calculation of the light conversion efficiency.

The design procedure for high rate ponds proposed by Oswald (1960) involves the following steps:

(1) Determine the concentration of algae (Equation 6-7)
(2) Compute the pond depth (Equation 6-6)
(3) Determine the light conversion efficiency, F
(4) Determine the solar energy, S, from prepared tables
(5) Compute the required detention period from Equation (6-5)

The application of this procedure requires a trial and error solution. The required tables can be found in the cited references.

TABLE 6-1. TYPICAL OXIDATION POND DESIGN CRITERIA
(After Oswald 1960)

	Anaerobic	Facultative	High rate
Depth, ft	8–10	2–5	0·6–1·0
Detention days	30–50	7–30	2–6
BOD loading, lb/acre per day	300–500	20–50	100–200
Per cent BOD removal	50–70	70–85	80–95
Algae concentration, p.p.m.	nil	10–50	>100

Aerated Stabilization Basins

A stabilization basin may be defined as a lagoon of significant depth in which aerobic conditions are maintained by recirculation or aeration with mechanical or diffused aeration equipment. The purification capacity of a stabilization basin is comparable to that in natural streams since the solids and turbulence levels are of the same order of magnitude. In a stabilization basin there is usually sufficient turbulence to create a condition of uniform concentration. It follows, therefore, that the concentration of BOD in the effluent is substantially equal to the concentration of BOD in the basin itself. Under these conditions, the removal efficiency may be expressed as follows when the removal approximates first-order kinetics (O'Connor and Eckenfelder, 1958):

$$\text{per cent } E = 100 \frac{K_1 t}{1 + K_1 t} \quad (6\text{-}9)$$

in which

$K_1 =$ BOD removal rate based on first order kinetics, 1/day
$t =$ basin retention period, days

The coefficient, K_1, is a measure of the rate of removal of BOD and may be determined experimentally in the laboratory under conditions simulating solids and turbulence levels which may be expected in the large-scale installation. Employing the reaction rates developed from laboratory data, the detention time and volume of the lagoon may be determined for a given removal in accordance with Equation (6-9).*

The oxygen requirements for a stabilization basin may be determined from the BOD loading and the active sludge solids, in accordance with Equation (2-14). In stabilization basins the sludge solids in suspension are negligible and the term bS in the equation may be neglected:

$$R_r = a'L_r \qquad (6\text{-}10)$$

in which

R_r = lb O_2/day required
L_r = lb BOD removed/day
a' = ratio between oxygen utilization and BOD removal

The constant, a', may be evaluated as shown in Chapter 2.

These oxygen requirements are supplied by the oxygen transferred from the atmosphere to the liquid contents of the basin through the water surface. Additional oxygen, if required, may be transferred by mechanical or diffused aeration systems. The total quantity of oxygen transferred may be defined by the following expression:

$$N = N_s + N_D \qquad (6\text{-}11)$$

in which

N = total quantity of oxygen transferred per unit time
N_S = quantity of oxygen transferred by surface aeration per unit time
N_D = quantity of oxygen transferred by diffused or mechanical aeration per unit time

In accordance with Equation (3-8), Equation (6-11) can be expressed:

$$N/V = \frac{dc}{dt} = K_2(C_s - C) + K'_L a(C'_s - C) \qquad (6\text{-}12)$$

or

$$N = K_L A(C_s - C) + K'_L a V(C'_s - C) \qquad (6\text{-}13)$$

* Equation (6-9) assumes first-order kinetics. It has been shown in Chapter 2 that in heterogenous waste mixtures the reaction rate, K_1, may decrease with time of aeration. In such cases Equation (6-9) must be modified for this decreasing reaction rate.

in which

$K'_L a$ = overall coefficient for diffused or mechanical aeration
V = basin volume
K_2 = overall coefficient of surface aeration
C = average dissolved oxygen concentration
K_L = liquid-film coefficient for surface aeration
A = surface area of lagoon
C_s = saturation value of oxygen at atmospheric pressure
C_s' = saturation value of oxygen for bubbles at the average depth of the lagoon
K_L' = liquid-film coefficient for mechanical or diffused aeration.

From the data currently available the concentration of oxygen in the liquid body, C, should be maintained above approximately 1·0 p.p.m. Below this level the uptake rate has been found to be oxygen concentration dependent. A stabilization basin exhibits a pronounced velocity gradient and the liquid film coefficient for surface aeration is expressed by Equation (3-14). The velocity gradient is induced by the fluid motion in the surface layers of the basin. This motion may be generated by diffused or mechanical aeration equipment. Depending upon the type and arrangement of the equipment, the supply of air could be a point or line source. In either case, the fluid motion and its associated turbulence characteristics decay exponentially with distance from the source. Figure 6-1 diagrammatically illustrates these cases. Typical vertical velocity profiles are also shown in this figure. In employing Equation (6-13) for design purposes, a mean value of the velocity gradient to calculate K_L must be used. The high velocity streamlines are found only in the upper layers moving away from the source, while the lower layer contains the low magnitude values in the reverse direction. It is pertinent to note that the surface area directly above the air source is subject to violent agitation and surface renewal is properly characterized by isotropic turbulence. This zone exists only in the immediate vicinity of the source and is not representative of the entire basin surface.

AEROBIC BIOLOGICAL TREATMENT PROCESSES

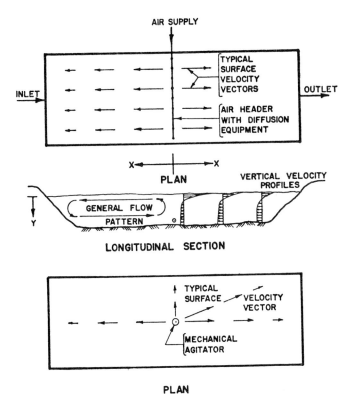

FIG. 6-1. Schematic representation of an aerated lagoon.

The decay of velocity with depth may be approximated by the exponential function as follows:

$$U = U_0 e^{-fy} \quad (6\text{-}14)$$

in which

U = liquid velocity at depth y
U_o = liquid velocity at surface
f = constant
y = depth

To obtain the velocity gradient at the surface, Equation (6-14) is differentiated with respect to y.

At the surface $U = U_0$ and:

$$\frac{dU}{dy} = -fU_o \qquad (6\text{-}15)$$

Equation (6-15) may be employed to compute the velocity gradient at any distance, x, from the aeration source. To calculate the total surface aeration, an average value of the velocity gradient over the entire lagoon surface is required. To this end, an empirical relationship developed from field data has been defined relating the velocity gradient with distance from a line source:

$$\frac{dU}{dy} = \frac{\phi}{x} \qquad (6\text{-}16)$$

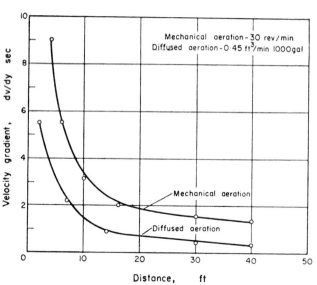

FIG. 6-2. Decay of velocity gradient with distance in an aerated lagoon.

This relationship is shown in Fig. 6-2. The average value of the velocity gradient may be obtained by integrating this function over the distance $x_1 - x_2$. The distance x_1 is selected as one foot from the source due to the indeterminate nature of the velocity gradient above the source. The distance x_2 is $\frac{1}{2}$ the spacing of the aeration units. The average velocity gradient computed from Equation (6-16) is substituted in Equation (3-14) to determine the average coefficient of surface reaeration. Measured velocity gradients have ranged from 0·1 to 4·0 per sec. These are of the same order as those found in shallow, turbulent streams.

Oxygen absorption from diffused or mechanical aeration is discussed in Chapter 3.

Under steady state conditions in a lagoon, the oxygen balance may be expressed by Equation (3-26).

$$K(C_s - C) = r_r$$

Equation (3-26) may be employed to determine the overall coefficient K, which measures the combined effect of surface and artificial aeration. r_r is measured in the field or laboratory and is related to the BOD removal.

The temperature effect on both the oxygen uptake and transfer rate has been discussed in Chapters 2 and 3, respectively. The overall temperature coefficient, θ, for the BOD removal was found to be 1·035 for this process. A temperature coefficient for aeration has been reported to vary from 1·016 to 1·047.

The addition of nutrients (nitrogen and phosphorus) increases the BOD removal rate in nutrient deficient wastes.

A portion of the BOD removed in the basin results in the synthesis of microbial sludge. A significant portion of these solids may settle to the bottom of the basin. In addition, solids present in the influent waste may also settle out. Sludge accumulation has been found to vary from 0·1–0·2 lb sludge/lb BOD removed for a pulp and paper mill waste.

Field and laboratory investigations have been conducted to evaluate the efficiency of this method of treatment. Treatment of cannery wastes showed an average BOD reduction of 43 per cent for one day detention to 73 per cent at 5 days without nutrients. The operating temperature was 25°C. Treatment of a pulp and paper waste at this temperature varied from 32 per cent reduction for one day to 70 per cent at 4 days without nutrients. Treatment of a board mill waste with nutrient addition showed 90% BOD reduction in 4–5 days detention. Typical laboratory and pilot plant data are shown in Fig. 6-3.

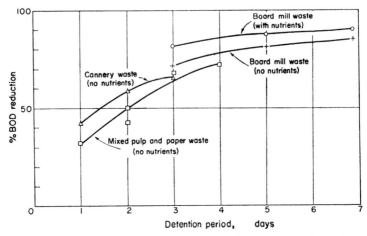

Fig. 6-3. BOD removal–time relationship for treatment of organic wastes in an aerated lagoon.

Example 6-1.

    Data   Flow   = 1 Mgal/day

              BOD raw = 250 p.p.m.

              Per cent Eff. = 70 per cent at Temp = 25°C

1. Volume of basin
   a. From laboratory data, the oxidation rate $K_1 = 0.53$ at 25°C (no nutrient addition)
   b. Detention time from re-expressing Equation (6-9)

$$t = \frac{E}{K_1(100-E)}$$

$$t = \frac{70}{0.53(100-70)}$$

$$t = 4.5 \text{ days}$$

AEROBIC BIOLOGICAL TREATMENT PROCESSES 201

c. Volume
$$V = 4.5 \text{ days} \times 1.0 \text{ Mgal/day}$$
$$V = 4.5 \text{ mg} = 600{,}000 \text{ ft}^3$$
assume 2 units each at 300,000 ft³

d. Dimensions
Each unit: assume depth = 10 ft
area = 30,000 ft²
length = 300, ft
width = 100 ft

2. Design of Aeration System
   a. From laboratory data, a' in Equation (6-10) was found to be 1·1;
   $$\#O_2/\text{day} = 1.1 \ \#\text{BOD}_r/\text{day}$$
   $$= 1.1 \times 0.70 \times 250 \times 8.34 \times 1.0$$
   $$= 1600 \ \#/\text{day} = 67 \ \#/\text{hr}$$
   b. A diffusion device was selected with an efficient operating range of 5 to 30 scfm per unit.

   For the 1st trial design, the following parameters were assumed:

   Air flow per unit = G_s = 15 scfm
   Air header spacing = 60 ft
   No. of headers per basin = 5
   D.O. concentration = C = 1·0 p.p.m.
   $C_s = 8.4$ p.p.m.
   $C'_s = 9.0$ p.p.m. at average depth

   c. The total oxygen transfer is expressed by Equation (6-13):
   $$N = K_L A(C_s - C) + K'_L a \ V(C'_s - C)$$

   Equation (6-13) may be re-expressed in terms of the number of diffusion units.
   $$N = K_L A(C_s - C) + nK'_L a \ V_0(C'_s - C)$$

From this equation, the number of diffusion units may be determined as follows.

Surface Aeration

Assume air flow = 0·4 scfm/1000 gal.

For this air flow, Equation (6-16) becomes:

$$\frac{dU}{dy} = \frac{11 \cdot 5}{x}$$

The mean velocity gradient is:

$$\begin{aligned}\frac{\overline{dU}}{dy} &= \frac{11 \cdot 5 \, (\log_e x)_1^{30}}{30 - 1} \\ &= \frac{11 \cdot 5 \times 3 \cdot 4}{29} \\ &= 1 \cdot 35/\text{sec} = 4900/\text{hr}\end{aligned}$$

(The upper limit of 30 ft is ½ the header spacing.)

The surface aeration coefficient may be now computed from Equation (3-14).

$$K_L = \left(D_L' \, \frac{\overline{dU}}{dy}\right)^{1/2}$$

for 25°C

$$\begin{aligned}D_L &= 0 \cdot 93 \times 10^{-4} \, \text{ft}^2/\text{hr} \\ K_L &= (0 \cdot 93 \times 10^{-4} \times 0 \cdot 49 \times 10^4)^{1/2} \\ &= 0 \cdot 67 \frac{\text{ft}}{\text{hr}}\end{aligned}$$

The surface area = 60,000 ft²

$$\begin{aligned}N_s &= K_L A (C_s - C) \\ &= 0 \cdot 67 \times 60{,}000 (8 \cdot 4 - 1 \cdot 0) \times 62 \cdot 4 \times 10^{-6} \\ &= 19 \, \#/\text{hr}\end{aligned}$$

e. Diffused Aeration

The required oxygen transfer from diffused aeration

$$N_D = N - N_s$$
$$= 67 - 19 = 48 \ \#/\text{hr}$$

The number of diffusion units may be determined:

$$N_D = nK_L'a\ V_0(C_s' - C)$$

in which
 $n =$ number of diffusion units
 $V_0 =$ lagoon volume/diffuser unit

for
 $G_s = 15$ scfm and a lagoon depth of 10 ft
 $K_L'aV_0$ for a particular diffuser $= 750$ ft^3/hr (see Chapter 3)

$$\therefore n = \frac{48}{750 \times (9 \cdot 0 - 1 \cdot 0) \times 62 \cdot 4 \times 10^{-6}} = 129$$

The number of diffusion units per header

$$= \frac{129}{5 \times 2} = 12 \cdot 9\ ;$$

use 13 and space at 100/13 or 7·7 ft.

f. Mechanical Aeration

For a given surface aerator $K_L a \cdot V \equiv 17{,}500$ ft^3/hr when operating at 30 rev/min.

Then as above,

$$n = \frac{48}{17{,}500 \cdot (8 \cdot 4 - 1) \cdot 62 \cdot 4 \times 10^{-6}} \equiv 5 \cdot 9 \text{ units}$$

use 6 units @ 100 ft spacing.

ACTIVATED SLUDGE

The activated sludge process may be defined as a system in which flocculated biological growths are continuously circulated and contacted with organic waste in the presence of oxygen. The oxygen

is usually supplied from air bubbles injected into the sludge–liquid mass under turbulent conditions. The process involves an aeration step followed by a solids–liquid separation step from which the separated sludge is recycled back for admixture with the waste. The aeration step may be considered in three functional phases:
  (a) a rapid absorption of waste substrate by the active sludge,
  (b) progressive oxidation and synthesis of the absorbed organics and organics concurrently removed from solution,
  (c) further aeration resulting in oxidation and dispersion of the sludge particles.

The kinetics of these phases are discussed in Chapter 2.

Various modifications of the activated sludge process have been developed to achieve economic advantage in construction and operation.

Conventional Activated Sludge

The conventional activated sludge process consists of four functional steps:
  (1) Primary sedimentation to remove settleable organic and inorganic solids.
  (2) Aeration of a mixture of waste and a biological active sludge.
  (3) Separation of the biologically active sludge from its associated treated liquor by sedimentation.
  (4) Return of settled biological sludge to be admixed with the raw wastes. A typical flow diagram is shown in Fig. 6-4.

For domestic sewage with normal *per capita* water consumption rates, most state health department standards require primary sedimentation for one hour, aeration detention capacity for 5–8 hr based on the average sewage flow plus 25 per cent return and the final sedimentation based on a surface overflow rate of 1000 gal/ft^2 per day. Conventional activated sludge treatment of domestic sewage has been shown to produce 90–95 per cent BOD reduction. Aerator loadings in conventional sewage treatment plants vary from 0·1 to 0·4 lb BOD/day/lb aeration sludge. (Haseltine, 1957). Waste sludge from conventional sewage treatment plants will generally vary from 10–30 per cent per day of the total aeration solids.

In the case of low BOD wastes (e.g. sewage, paper mill wastes, etc.) the process usually operates over the range B–C on Fig. 6-5.

Figure 6-5 is detailed in accordance with the definitions of Chapter 2. This requires an aeration detention period of t_1 and results in a sludge accumulation of Δs_1. The mathematical relationships governing BOD removal under these conditions are shown by Equations 2-8 through 2-11. A major part of the BOD is removed in the first

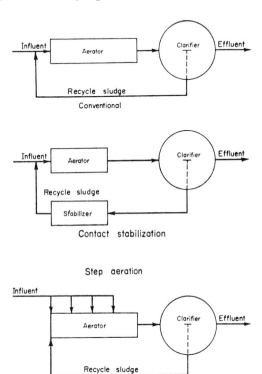

FIG. 6-4. Activated sludge flow diagrams.

few minutes of aeration by biosorption and flocculation. Oxidation and synthesis occur during the remainder of the aeration period. Since oxygen utilization is intimately associated with BOD removal and sludge growth, the oxygen demand rate curve will approximately follow the BOD removal curve. In the conventional process this results in a high initial demand rate followed by a rapid decrease to the endogenous rate toward the end of the aeration period. For average sewage strengths the oxygen uptake rate will vary from about 60 p.p.m./hr at the head end of the aeration tank to approximately 15 p.p.m./hr at the outlet end. Tapered aeration is necessary to follow this demand rate curve. Process modifications such as step aeration

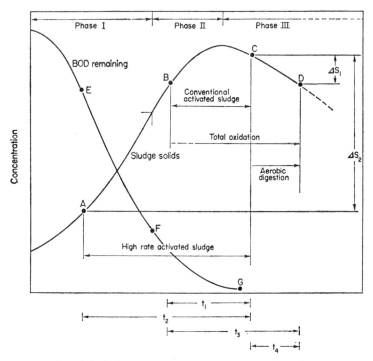

FIG. 6-5. Sludge growth and BOD removal relationships for various activated sludge processes.

in which the sewage is added to the aeration tank in increments down its length or complete mixing in which the sewage is added along the length of the aeration tank will equalize the oxygen demand throughout the aeration tank at its mean rate. This will insure more uniform distribution of oxygen and maintenance of dissolved oxygen.

Mixed pulp and paper mill wastes have been treated by the conventional process at the West Virginia Pulp and Paper Company,

Covington, Va. (Laws and Burns, 1960) for several years. Recent operating data is summarized below:

AVERAGE PERFORMANCE FOR NOV.–JAN., 1960
ACTIVATED SLUDGE PLANT—WEST VIRGINIA PULP & PAPER
CO., COVINGTON, VIRGINIA

Flow, Mgal/day	21·0
BOD, p.p.m.	188
Suspended solids, p.p.m.	122
Lb. BOD applied/per day/lb aeration sludge	0·67
Mixed liquor suspended solids, p.p.m.	2280
Sludge Volume Index	286
Recycle sludge (per cent)	30
Aeration detention, hr	2·3
Per cent BOD removal	84
Lb N/100 lb BOD removed	3·4
Lb P/100 lb BOD removed	0·7

BOD removal from this waste follows a retardent reaction whereby 67–75 per cent of the BOD is rapidly removed from solution in the first hour of aeration and the remainder at a progressively decreasing rate. BOD removals in excess of 85–90 per cent consequently require long aeration detention periods. The mean oxygen utilization was found to vary from 35·3 p.p.m./hr per 1000 p.p.m. sludge at the influent end of the aeration tank to 8·9 p.p.m./hr per 1000 p.p.m. sludge at the effluent end.

In the case of high BOD wastes (BOD > 500 p.p.m.), the process operates over the range A–C (Fig. 6-5). For high BOD removals the detention period is increased to t_2 and the excess sludge accumulation to Δs_2. Down to a limiting concentration of BOD, the BOD removal is approximately linear with sludge concentration and time (E–F) Fig. 6-5: Equations 2-7b and 2-7c. Below this BOD concentration, the rate of removal is concentration dependent.

The oxygen utilization rate will remain at a constant or a slightly increasing value over the range E–F. Below point F, Fig. 6-5, the uptake rate will rapidly decrease and level out at the endogenous rate. In the treatment of a pharmaceutical waste (Dryden et al., 1956), the BOD removal rate averaged 200 p.p.m./hr per 1000 p.p.m. sludge to a limiting concentration of approximately 400 p.p.m. At lower concentrations the removal rate progressively decreased. Typical conventional activated sludge treatment data is shown in Table 6-2.

TABLE 6-2. AVERAGE PERFORMANCE DATA FOR CONVENTIONAL ACTIVATED SLUDGE PLANTS

Waste	BOD raw waste p.p.m.	% BOD reduction	% return sludge	Aeration detention hr	Mixed liquor solids p.p.m.	Oxygen utilization p.p.m./hr	Temp. °C	BOD loading lb BOD/day per lb sludge
Sewage[1]	182	89·6	32·0	5·3	1844	—		0·34
	217	93·6	21·5	9·27	937	—		0·50
	97	92·0	38·5	3·3	2330	—		0·22
	254	87·0	31·5	6·7	1808	—		0·39
	193	94·8	83·3	8·9	2800	—		0·10
	125	90·0	20·8	4·8	1200	—		0·44
Pulp and paper	183	92·0	33·0	3·8	2910	33·0	25	0·30
Cannery[2]	1350	91·5	—	4·8	3000	44·5	—	
Pharmaceutical[3]	1684	91·0	100·0	2·5	3900	108·0	20	2·07
	2532	57·0	33·0	2·5	1890	143·0	20	9·70
Pulp and paper	446	91·0	53·0	10·2	2103	17·1	32	0·43
Kraft mill	144	85·0	—	8·0	900	—	25	0·48
Ammonia still	100	93·0	46	5·0	2530	—	—	0·31
Refinery	1100	87·0	—	4·0	7700	—	—	1·14
White water[4]	2000	86·0	—	8·0	2000	—	20	3·0

[1] HASELTINE (1956).
[2] ECKENFELDER and GRICH (1955).
[3] DRYDEN et al. (1956).
[4] RUDOLFS and AMBERG (1953).

Dispersed Growth Aeration

Heukelekian (1949) has shown that in certain cases difficulty due to dispersing or bulking floc is encountered in the treatment of strong industrial wastes by the activated sludge process. Treatment by dispersed growth aeration or non-flocculent growth consists of innoculation of the waste with a sewage, soil or other suitable bacterial suspension and aeration. Acclimatization is accomplished by gradually decreasing the quantity of seed added daily and increasing proportionally the quantity of returned supernatant seed. Acclimatization is usually continued for a week. An active growth is obtained in this manner.

At the end of the aeration period any settleable solids produced are settled and removed and a portion of the settled liquor returned for seeding the incoming waste. BOD reduction depends on the applied BOD and the time of aeration. Heukelekian (1949) found 80 per cent BOD reduction in 24 hr aeration with wastes with an initial BOD up to 3000 p.p.m. Ninety per cent BOD reduction was obtained with wastes containing less than 1000 p.p.m. BOD. Nemerow and Rudolfs (1952) on rag, rope and jute wastes found reductions of 60–70 per cent in 24 hr aeration at 20°C. The feed waste BOD was 2000 p.p.m.

Contact Stabilization

When a high percentage of the BOD is rapidly removed by biosorption after contact with well aerated activated sludge the contact-stabilization process can be advantageously employed. In this process, waste is aerated with stabilized sludge for a short contact period (usually 30–60 min). The mixed liquor is then separated by sedimentation. The settled sludge is transferred from the clarifier to a sludge stabilizer where aeration is continued to complete the oxidation and to prepare the sludge for BOD removal of fresh incoming waste. The process is schematically illustrated in Fig. 6-6.

When the BOD removal rate is too low to attain the desired overall BOD removal in a short contact period, the aeration contact period can be extended to attain the additional desired removal. This additional removal is indicated in Fig. 6-6.

The magnitude of the initial removal is dependent on the condition of the sludge when it is in contact with waste and on the waste characteristics. The relation between sludge aeration or stabilization time and BOD removal for various contact periods is shown in Fig. 6-7. The data in Fig. 6-7 was obtained by aerating sludge with a mixture of domestic sewage and textile mill waste for 15 min, separating the sludge, and aerating the sludge with waste for 0·25, 1·0 and 1·5 hr after the indicated stabilization periods. It is apparent from Fig. 6-7 that for a contact period of 0·25 hr, a stabilization period of 2–6 hr is necessary to maintain the BOD removal capacity of the

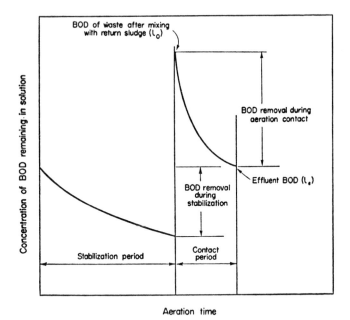

Fig. 6-6. The contact-stabilization process.

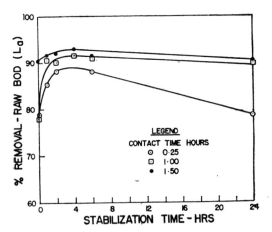

Fig. 6-7. Effect of stabilization time on BOD removal during various contact periods.

sludge. Increasing the contact period to one hour permits some oxidation of the absorbed BOD and reduces the stabilization requirement to one hour.

The required stabilization period will depend on the magnitude of initial BOD removal and on the contact detention period. In-

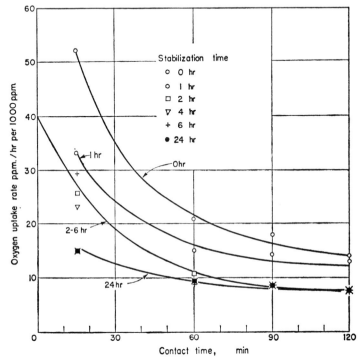

FIG. 6-8. Oxygen utilization characteristics of the contact stabilization process.

creasing the contact period permits some oxidation of absorbed organics and reduces the required stabilization period. A substantial increase in the contact period will permit complete oxidation of the absorbed organics and eliminate the need for stabilization. This is then a conventional activated sludge process. Increasing the concentration of mixed liquor suspended solids will decrease stabilization requirements for the same BOD loading since the BOD removal per unit sludge mass will be reduced. Under these conditions less time is required for the oxidation of the absorbed BOD in the stabilization tank. The variation of oxygen utilization rate for various contact and stabilization times is shown in Fig. 6-8.

The contact stabilization process has been employed for the treatment of domestic sewage at Austin, Texas, by Ullrich and Smith (1951) and Bergen County, New Jersey, by Zablatsky (1959). Pilot plant and full-scale plant studies were reported employing this

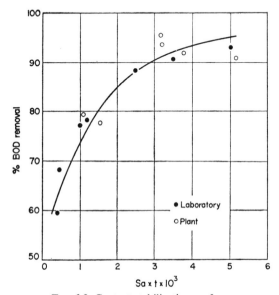

FIG. 6-9. Contact-stabilization performance.

ocess for the treatment of cannery wastes, by Eckenfelder and Grich (1955). High BOD reductions have been reported in the application of this process to the treatment of pulp and paper mill wastes (Rice and Weston, 1951). BOD reductions in excess of 85 per cent have been found in the treatment of a mixture of domestic sewage and textile wastes (Cone Mills, 1959). The performance of the contact stabilization process under various operating conditions is shown in Fig. 6-9. The performance of the contact-stabilization process treating various wastes is summarized in Table 6-3.

TABLE 6-3. PERFORMANCE OF THE CONTACT-STABILIZATION PROCESS

Waste	BOD	Detention, hr		Suspended solids, p.p.m.		% Re-cycle	% BOD reduction
		Contact	Stabilization	Contact	Stabilization		
Sewage	264	0·24	1·60	3251	5218	100	92·5
Sewage[1]	108	1·30	4·25	2239	8629	44	88·0
Sewage[1]	210	0·18	1·00	2500	4500	100	90·0
Sewage and textile mill[2]	225	0·60	5·50	2950	6950	67	77·0
Sewage and textile mill[2]	320	1·10	3·30	3200	7900	71	86·0
Pulp and paper	249	0·50	3·70	6360	—	25	72·0
Pulp and paper	191	1·00	4·0	4642	—	100	87·0
Pulp and paper	218	2·00	2·0	5980	—	25	93·0
Tomato cannery[3]	412	0·80	1·6	2250	3600	100	85·0
Tomato cannery[4, 3]	450	0·35	1·65	2250	4500	125	84·0
Tomato and apple[3]	492	1·00	2·00	2500	4400	100	89·7
Peach and tomato[3]	740	0·65	1·30	3600	5900	100	58·0

[1] ZABLATSKY et al. (1959).
[2] CONE MILLS (1959).
[3] ECKENFELDER and GRICH (1955).
[4] Aero-cyclator stabilizer pilot plant.

Step Aeration

Step aeration involves the addition of primary sewage effluent along the course of flow of mixed liquor through the aeration tanks to equalize the loading and the oxygen demand rates through the aeration tanks. The return sludge is admitted to the aeration tank influent and the primary effluent added at multiple points along the aerator by means of stop gates. A flow diagram is shown in Fig. 6-4. Aeration detention periods are generally of the order of 2–4 hr. Return sludge rates are generally maintained at 25–35 per cent of the average sewage flow. Operating results compare favorably with conventional treatment.

Modified Aeration

Intermediate treatment is attained by modified aeration in which the aeration detention period is about 2 hr based on the average

TABLE 6-4. AVERAGE PERFORMANCE DATA FOR PLAIN AERATION AND HIGH RATE ACTIVATED SLUDGE

Waste	BOD raw waste p.p.m.	% BOD red.	% re-cycle	Aeration detention hr	Mixed liquor solids p.p.m.	Temp. °C	BOD loading BOD/day per lb sludge
Sewage	171	63·1	0·0	6·6	150	—	4·15
	132	58·4	24·0	3·9	522	—	1·26
	170	58·0	10·6	2·3	360	—	4·65
	103	76·7	28·5	2·5	505	—	1·68
Rag, rope and jute	2000	65	—	24·0	—	20°C	—
White water	485	80·0	100	4·3	—	30°C	—

sewage flow plus 10 per cent return sludge. Low mixed liquor solids (200–500 p.p.m.) and a high loading factor (low sludge age) is maintained. BOD removal efficiencies are of the order of 50–70 per cent in domestic sewage treatment. The flow diagram is substantially similar to the conventional process. Some typical operating data are shown in Table 6.4.

Aero-Accelator

A unit has been developed which combines the features of aeration and clarification in one structure (Busch and Kalinske, 1955). In this unit, known as the Aero-Accelator, the raw waste enters the bottom of the unit where it is admixed with recirculating sludge and air. The mixture is aerated and dispersed for intimate contact with a large volume of activated slurry beneath the hood section from which it passes vertically upward in the center of the unit and then flows radially through adjustable ports. The sludge is separated from the treated liquor and recirculated at the bottom of the unit. Aeration is accomplished by impinging a flow of air from an open pipe against a rotating agitator.

The Aero-Accelator contains an inner and an outer draft tube which encompasses approximately two-thirds of the total tank volume and through which the sludge is recirculated at a high rate.

Oxygen transfer through gas-liquid contacting occurs only in the draft tube composing approximately one-third of the total tank volume. Biological oxidation is considered to occur throughout both draft tubes due to the high interchange of oxygen and sludge between the two tubes. Process data from Aero-Accelator operation on packing-house, refinery, pulp and paper and domestic wastes are shown in Table 6-5. This unit is shown in Fig. 6-10.

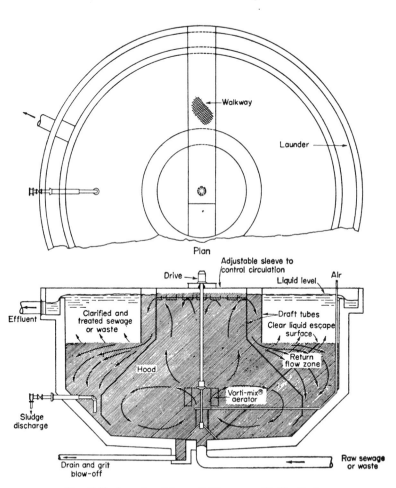

FIG. 6-10. Aero-Accelator unit (*Courtesy of Infilco Inc.*).

TABLE 6-5. AVERAGE PERFORMANCE DATA FROM AERO-ACCELATOR PLANTS

Waste	BOD Loading lb/day per lb. M.L. S.S.	Raw BOD p.p.m.	M.L. S.S. p.p.m.	Aeration Time hr	BOD Reduction %
Soluble organics	0·41	1,000	15,000	2·1	90
Pea and carrot	2·5	1,500	2,500	3·1	95
Beet canning	2·0	4,000	2,600	9·3	96
Phenol	0·18	200	2,000	7·0	92
Pulp and paper	0·48	188	3,200	1·6	89·5
Pulp and paper	0·57	600	4,200	3·1	93
Packing house waste	1·36	960	3,500	2·5	88
Sewage	0·78	330	3,800	1·4	89
Naval base sewage	0·56	250	4,800	1·3	92
Sewage	0·73	320	2,500	2·3	92
Sewage	0·80	200	3,000	1·0	90
Sewage	0·90	118	2,100	0·85	90
Sewage	0·85	275	2,200	2·20	83
Sewage	1·10	130	2,000	0·75	85
Sewage	0·55	175	4,000	1·1	75

Total Oxidation

Total oxidation, by definition, is a process so designed that the biological sludge produced by synthesis is consumed by auto-oxidation. In order to accomplish this oxidation, the aeration detention period must be increased from t_1 to t_3 (Fig. 6-5). The process then operates over the range B–D and the sludge accumulation, ΔS, theoretically approaches zero.

In practice, it has been shown that the auto-oxidation of biological sludge does not follow-first order kinetics, but rather follows a retardent reaction in which the rate of oxidation decreases with time or concentration. This is due to the fact that the various cellular constituents differ in their ease of oxidation. A portion of the cellular material is highly resistant to oxidation and results in an accumulation in the process. McKinney (1960) and Kountz *et al.* (1959) have indicated that the non-oxidizable solids buildup from the auto-oxidation of biological sludge may amount to 25 per cent of the sludge formed. If non-oxidizable suspended solids are present in the incoming waste the solids buildup will be greater. In small installations, this excess sludge accumulation can be removed at intermittent intervals by tank truck.

The total oxidation process can either be designed as a continuous system in which the aeration detention period is usually 24 hr or as

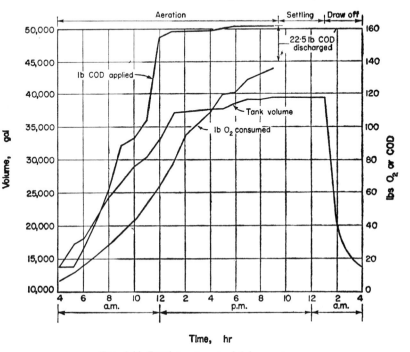

FIG. 6-11. Batch treatment of dairy wastes.

a batch fill and draw system. The batch system is most adaptable to industrial wastes or sewages in which the total waste flow is received over 8–12 hr. The operation of a batch system involves aeration, settling and decantation from a single tank. This sequence of operations for a dairy waste is shown in Fig. 6-11.

Since the detention period required for BOD removal is considerably less than that required for auto-oxidation the tank volume will be controlled by the sludge oxidation rate. The following procedure can be employed to determine design criteria.

The sludge balance in the aeration tank can be approximately expressed by the relationship:

$$\Delta S = aL_r - bS_a \tag{6-17}$$

in which ΔS = increase in biological sludge solids in lb/day

a = fraction of BOD synthesized to sludge. This usually varies from 50–75 per cent of the 5-day BOD removed

L_r = BOD removed, lb/day

b = mean rate of auto-oxidation, fraction per day. For soluble substrates such as dairy wastes, b will average between 15–25 per cent per day at 25°C. Wastes containing suspended solids of a low oxidation rate such as domestic sewage may have endogenous rates of 6–12 per cent per day.

S_a = average mixed liquor suspended solids, lb.

Equation (6-17) does not consider the inert fraction of sludge which is not oxidized in the process. In the total oxidation process $\Delta S \simeq 0$ and Equation (6-17) becomes:

$$aL_r = bS_a \tag{6-18}$$

The quantity of sludge which must be maintained under aeration in order to approach total oxidation is:

$$S_a = aL_r/b \tag{6-18a}$$

The application of Equation (6-18a) can be illustrated by the following example:

Example 6-2. Design a total oxidation plant to treat 26,000 gal/day of a dairy waste containing 161 lb COD over a period of 17 hr. The mean auto-oxidation rate of the sludge is 24 per cent per day. The COD reduction is 90 per cent.

$$\text{COD removed/day} = 0.90 \times 161 = 145 \text{ lb/day}$$

Equilibrium quantity of sludge

$$\Delta S_a = aL_r - bS_a$$
$$\Delta S_a \cong 0$$
$$aL_r = bS_a$$

If the constant a for dairy waste is 0·5 and if aeration proceeds for 17 hr then

$$(0\cdot5)\,(145) = 0.24 \cdot \frac{17}{24} S_a$$

$$S_a = 430 \text{ lb}$$

If the sludge occupies 0·8 ft³/lb:

The required sludge volume is (0·8) (430) =	344 ft³
The waste volume is 26,000/7·48	= 3500 ft³
Sludge free board	= 172 ft³
Total tank volume	= 4016 ft³

Since in the total oxidation process the sludge produced by synthesis is consumed by auto-oxidation, the oxygen requirements are greater than the conventional process. For the case of complete sludge oxidation the total oxygen requirements will approximately equal the ultimate BOD.

Activated Sludge Design Formulations

The degree of treatment attainable in the activated sludge process is functionally dependent on the size of the aeration basins and the active solids which can be carried in the aeration tanks, providing sufficient oxygen is supplied to the process. The mixed liquor solids concentration is in turn limited by the settling and compaction characteristics of the biological sludge in secondary settling tanks. It is frequently convenient to relate BOD removal efficiency to a loading parameter expressed as lb BOD applied per day per lb of aeration sludge.

The loading factor employs mixed liquor suspended solids for convenience. Since only a portion of the sludge may be considered as active culture, observed removals for particular systems are only representative of that specific waste. In some cases volatile solids will provide a better correlation than total solids. An example from a pulp and paper waste oxidation process may be cited. The sludge was 85 per cent volatile but due to the presence of biologically inert fibre and other volatile solids the computed active fraction was only 70 per cent. This active fraction was variable, depending upon the detention time and fibre load on the process.

The loading applied to the sludge is related to various secondary factors which influence the overall process performance. Many investigators have interpreted these effects in terms of sludge age which is related to the length of time the sludge has been undergoing aeration. Sludge age may generally be considered as the reciprocal of the loading factor. Gould (1953) has defined sludge age as the mixed liquor suspended solids divided by suspended solids per day in the raw sewage. A more fundamental interpretation of sludge age was advanced by Gellman and Heukelekian (1953) as the mixed liquor suspended solids divided by lb BOD per day removed in the system.

While there is some doubt as to interpretation of the effects of high loadings on the overall process performance, it is generally conceded that sedimentation and compaction of sludge is impaired when high loading factors are employed. These factors are more fully discussed in a preceding section.

The design formulation can be summarized:

$$\text{loading factor} = \frac{\text{lb BOD applied/day}}{\text{lb MLVSS}} = \frac{24\, la}{S_a t(1+r)}$$

(6-19)

in which

la = p.p.m. BOD applied

S_a = average MLVSS

t = aeration detention period

r = sludge recirculation ratio

The loading factor can be plotted against the per cent BOD removal to obtain a generalized relationship. For specific cases, more

FIG. 6-12a. Biological filtration units. (*Courtesy of Dorr-Oliver Inc.*)

Fig. 6–12b. Biological filtration units.
(*Courtsey of Dorr-Oliver Inc.*)

AEROBIC BIOLOGICAL TREATMENT PROCESSES. 221

exact relationships can be developed, as shown in Chapter 2, based on sludge growth relationships. For high BOD loading levels Equations (2-7b) and (2-7c) can be employed. For low BOD loading levels Equations (2-8) to (2-11) can be employed.

Example 6-3. From pilot plant data it was found that 90 per cent BOD removal efficiency can be obtained with a loading factor of 0·9. If the applied BOD concentration is 263 p.p.m. in a flow of 5·7 Mgal/day, compute the required aeration tank volume.

(Settled sludge return = 10,000 p.p.m.; MLVSS = 2100 p.p.m. and MLSS = 2500 p.p.m.)

From a material balance of the process (neglecting solids in the influent waste)

$$r = \frac{S_a}{S_r - S_a} = \frac{2500}{10,000 - 2500}$$

Recycle flow $(R) = 0\cdot 33 \cdot 5\cdot 7$ Mgal/day $= 1\cdot 9$ Mgal/day
Solving for t
lb BOD_A/day per lb sludge $= 0\cdot 9$

$$t = \frac{24 \cdot 263}{2100 \cdot 0\cdot 9(1 + 0\cdot 33)} = 2\cdot 5 \text{ hr}$$

The total flow is: $Q + R = 5\cdot 7 + 1\cdot 9 = 7\cdot 6$ Mgal/day.
The aeration tank volume is:

$$7\cdot 6 \text{ Mgal/day} \cdot \frac{1}{24} \cdot 2\cdot 5 = 0\cdot 79 \text{ mg}$$

TRICKLING FILTERS

The trickling filter process may be defined as a fixed bed system over which sewage or waste is intermittently or continuously discharged and contacted with biological films on the filter media. Aerobic conditions are maintained by a flow of air through the filter bed induced by the difference in specific weights of the atmosphere inside and outside the bed. In some cases forced draft ventilation is employed. Periodic sloughing of filter film is discharged to secondary clarification units. The trickling filter is illustrated in Fig. 6-12.

Trickling filters may be broadly classified into high rate and low rate systems. The distinction between these systems is usually based on the hydraulic and organic loading to the filter. Low rate filters operate with intermittent dosing at hydraulic loadings of 2–6 Mgal/acre/day with organic loadings of 1500–7500 lb per acre-ft per day. The depth most commonly employed is 6 ft. High rate filters employ recirculation in single or two-stage units. A portion of the filter effluent is recirculated to the primary clarifier or the filter influent. Hydraulic loadings will usually vary from 10–30 Mgal/acre/day, although recent filter designs have employed hydraulic loadings

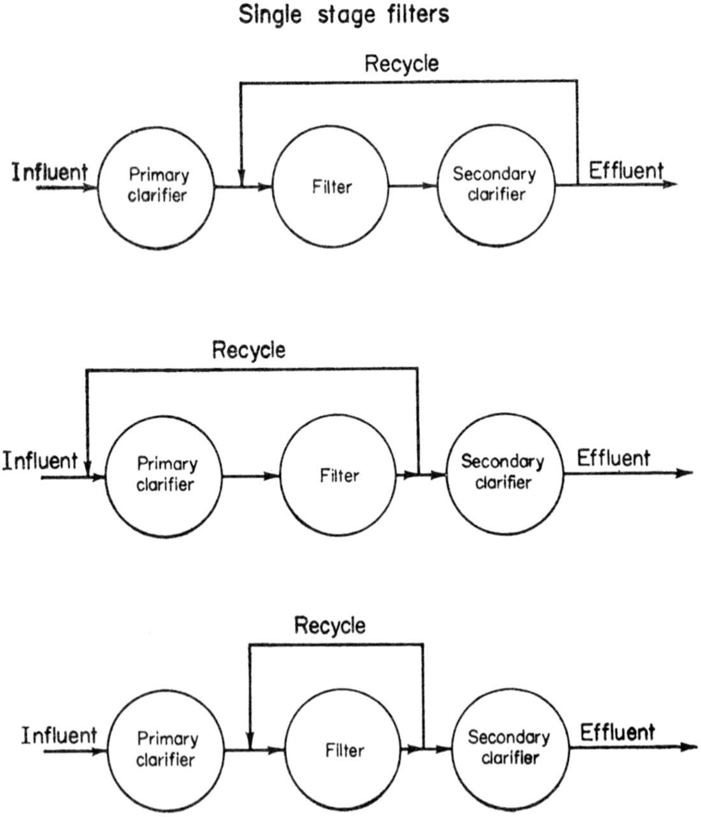

Fig. 6-13. Biological filtration flow sheets.

AEROBIC BIOLOGICAL TREATMENT PROCESSES   223

as high as 100 Mgal/acre/day. Some typical filter flow sheets are shown in Fig. 6-13.

The quantity of film on the filter contact surfaces will vary with the depth of the bed and with seasonal temperature. In a low rate filter the quantity of film has been found to vary from 8–12 lb/yd^3 of bed volume while in a high rate filter variations of 5·5–11 lb/yd^3 have

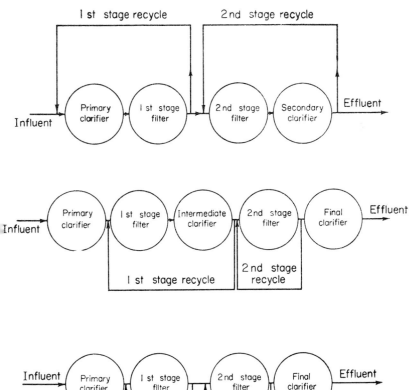

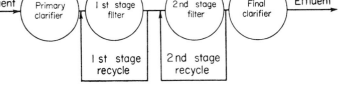

FIG. 6-13. Biological filtration flow sheets.

been found (Heukelekian, 1945). In polystyrene filter media developed by the Dow Chemical Co., 4·6 lb of sludge was found per yd^3 of 70 per cent volatile content. The film varied from $\frac{1}{16}$ in. to $\frac{1}{8}$ in. thickness. The BOD removal efficiency is not directly related to the total quantity of film present but rather to the portion of the film serving as an active oxidation device. This amount of film will vary with the applied filter loading and the temperature.

The Mechanism of BOD Removal in Trickling Filters

The mechanism of BOD removal in a trickling filter is similar to that of the activated sludge process. A large portion of the liquid applied to the surface of a filter passes rapidly through the filter and the remainder slowly trickles over the surface of the slime growth. Removal occurs by biosorption and coagulation from that portion of the flow which passes rapidly through the filter and by the progressive removal of soluble constituents from that portion of the flow with increased residence time. Since the residence time of liquid in the filter is related to the hydraulic loading it can be expected that the total removal of BOD will be related to the hydraulic loading. At high hydraulic loadings the fraction of applied BOD which can be removed in the filter is approximately equal to the maximum removal of BOD which can be attained by absorption in the activated sludge process and will depend on the chemical nature of the organic matter constituting the BOD. At low hydraulic loadings the increased contact time permits increased BOD removals in a manner analogous to increasing the contact time in the activated sludge process. The mechanism of BOD removal in a filter is shown in Fig. 6-14. At the low BOD concentrations usually applied to a filter, the fraction of BOD remaining is directly related to contact time (assuming first-order kinetics) (Howland, 1958) (Shulze, 1960)

$$\frac{Le}{L_0} = e^{-Kt} \qquad (6\text{-}20)$$

in which

Le = BOD remaining in filter effluent

L_0 = BOD applied to the filter after mixing with recycle

AEROBIC BIOLOGICAL TREATMENT PROCESSES

in which K is a function of the active film per unit volume. The contact time, t, is related to the depth and hydraulic loading as well as to the physical characteristics of the filter media. For sewage and complex industrial wastes, Equation (6-20) may require modification to a retardent function.

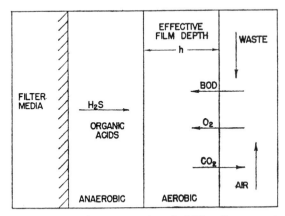

FIG. 6-14. Schematic representation of trickling filter operation.

It has been shown that the mean contact time through filter media can be expressed by the relationship: (Howland, 1958) (Shulze 1957)

$$t = CD/Q^n \qquad (6-21)$$

in which

t = contact time

D = filter depth

Q = hydraulic loading

The constant, C, and the exponent, n, incorporates surface and viscosity effects and will vary with the type of filter media. Theoretical analysis indicates that the exponent, n, will approach 1/3 for turbulent flow and 2/3 for laminar flow (Howland *et al.* 1960).

Howland (1958) developed a relationship for the residence time of liquid flowing over a sphere:

$$t = 2\cdot 6 \left(\frac{3v}{g}\right)^{1/3} \frac{2\pi^{2/3} r^{5/3}}{Q^{2/3}} \qquad (6\text{-}22a)$$

in which v = kinematic viscosity
r = radius of the sphere
g = gravitational constant

Equation (6-22a) is a specific form of Equation (6-21). The contact time for flow down an inclined plane was shown by Howland (1958) to be

$$t = \left(\frac{3v}{Sg}\right)^{1/3} \frac{A}{W^{4/3} Q^{2/3}} \qquad (6\text{-}22b)$$

W = width of plane

in which A = area
S = slope of inclined plane

McDermott (1957) determined the contact time of synthetic sewage passing over a 23 ft column of $3\frac{1}{2}$ in. balls as:

$$t = 30\cdot 2 \; D^{1\cdot 08} \; Q^{-0\cdot 55} \qquad (6\text{-}22c)$$

in which t is the contact time in seconds. Sinkoff et al. (1959) employed dimensional analysis to develop relationships for residence time. They obtained a generalized relationship for residence time.

$$t = \frac{CDv^a}{g^{1/3}} \cdot \left(\frac{S}{Q}\right)^b \qquad (6\text{-}23)$$

in which S is the specific surface and D the filter depth. The exponents a and b were found to be 0·5 and 0·83 and 0·09 and 0·42 for glass spheres and porcelain spheres respectively. Modifying Howland's equation to the Sinkoff form yields exponents a and b of 1/2 and 2/3 respectively.

The data of Sinkoff et al. (1959) indicated a variation in mean residence time of 7·5 min to 0·5 min over a range of hydraulic loading of 8·9 Mgal/acre/day to 276 Mgal/acre/day. By contrast, studies by the Water Pollution Research (1952) showed that at low hydraulic loadings (3–5 Mgal/acre/day) average residence times of 40–60 min were obtained.

Studies by Schulze (1960) on a screen filter showed residence time to be inversely proportional to $Q^{0.66}$. A study of residence time in Dowpac Plastic media (Bryan and Moeller, 1960) indicated residence time to be inversely proportional to $Q^{0.5}$.

The presence of slime increases the residence time. The residence time is also increased due to capillary water storage (Howland et al., 1960). With small size filter media and low hydraulic loading this storage can be as much as 100 per cent of the hydraulic loading.

It should be noted that most filters are intermittently dosed and therefore the mean application rate is not representative of the actual contact time of the sewage through the filter. Gutierezz (1955) has shown that slime growth can be developed and maintained at instantaneous application rates in excess of 300 Mgal/acre/day.

It has been previous shown that BOD is rapidly removed from a waste by agglomeration and coagulation and biosorption. Soluble constituents are removed more slowly on contact of waste with slime. The effect of residence time in a filter on BOD removal has been described by Howland (1958) and Shulze (1960) and is defined by Equation (6-20). The fraction of BOD remaining in solution can be approximately related to the hydraulic loading by the relationship:

$$\frac{Le}{L_0} = CQ^n \qquad (6\text{-}24)$$

A general relationship can be developed by combining Equations (6-20) and (6-21) with modification for slime distribution in the filter:

$$\frac{Le}{L_0} = e^{-KD^m/Q^n} \qquad (6\text{-}25)$$

Howland (1958) and Shulze (1960) have shown that m equals 1·0 in cases where the film is approximately uniformly distributed through the filter depth. In most installations, however, the film distribution, activity and composition vary with filter depth, being greatest in the surface layers. This condition results in the exponent m being less than 1·0.

Equation (6-25) indicates that at any given organic loading, an increase in hydraulic loading will result in a decrease in removal efficiency. This observation was also made by Velz (1948) and Grantham *et al.* (1950). Shulze (1960) determined K to be 0·3 and n to be 0·66 on a screen filter treating settled sewage. The results of McDermott (1957) defined K as 0·285. McDermott's results are summarized in Fig. 6-15.

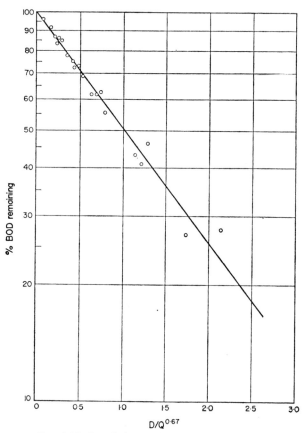

Fig. 6-15. Correlation of biological filtration data (after McDermott, 1957).

A unit volume of filter will have a limiting capacity to assimilate BOD into cell material and respiratory products. When this limiting capacity is exceeded, no further removal can occur. The limiting capacity is related to the temperature, the quantity of active film present and the oxygen transfer mechanism through the filter.

Equation (6-25) presumes that all components of the organic waste are removed at the same rate. There is considerable evidence, however, that in complex wastes the removal rate decreases with concentration or time since the more readily assimilable components will be removed more rapidly. This requires that a retardant form of equation be employed to describe the overall removal process.

Such a form of an equation which has been applied to filtration data is:

$$\frac{L_e}{L_0} = \frac{1}{1 + C' D^m / Q^n} \tag{6-25a}$$

Trickling Filter Design

Equation (6-25a) or a modification to consider the retardant effect on BOD removal may be employed for filter design. Published data from various filter installations treating domestic sewage are plotted in Fig. 6-16. The exponent 0·67 on the filter depth, D, was developed from a multiple correlation of published data. The exponent 0.50 on the hydraulic loading, Q, was developed from the data of Keefer and Meisel (1952) and the NRC report (1946). The retardant effect is observed by the decreasing slope of the removal curve with increased depth or hydraulic loading. Results obtained by Keefer and Meisel (1952) at Baltimore also indicated that BOD removal from settled sewage followed a retardant reaction. Their results are shown in Fig. 6-17.

The application of Fig. 6-16 to the design of a filter treating settled domestic sewage without recirculation is illustrated by the following examples:

Example 6-4

Compute the BOD removal from settled sewage in a 6 ft filter with a hydraulic loading of 12·0 Mgal/acre/day without recirculation.

From Fig. 6-16, for

$$\frac{D^{0.67}}{Q^{0.50}}$$

of 0·96, the per cent BOD remaining is 29 per cent. The BOD removal through the filter is therefore 71 per cent.

Example 6-5

Design a filter to yield 85 per cent BOD removal from settled sewage without recirculation.

From Fig. 6-16

$$\frac{D^{0.67}}{Q^{0.50}} = 2.27 \quad \text{for 15 per cent BOD remaining.}$$

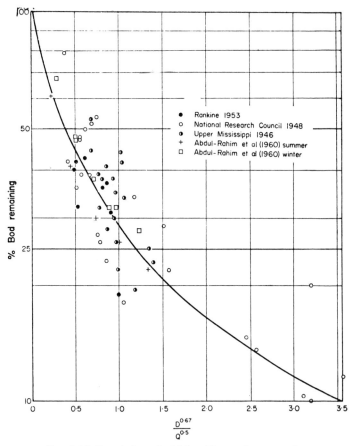

FIG. 6-16. Correlation of trickling filter performance data.

Using a depth of 6·0 ft, the allowable hydraulic loading is

$$\frac{D^{0.67}}{Q^{0.50}} = 2\cdot 27 \qquad Q = 2\cdot 13 \text{ Mgal/acre/day}$$

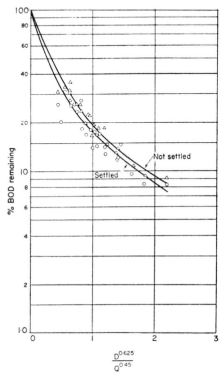

Fig. 6-17. Correlation of biological filtration results from Baltimore (after Keefer and Meisel, 1952).

Fairall (1956) statistically analyzed performance data from several trickling filtration installations in the western part of the United States. For filters with recirculation he developed the formulation:

$$\frac{Le}{L_0} = 1\cdot 102 \left(\frac{Q}{D}\right)^{0.322} \qquad (6\text{-}26)$$

Several other design formulations have been developed for trickling filters treating domestic sewage.

Velz (1948) showed that each unit of filter depth will remove a constant fraction of removable BOD applied to that unit depth:

$$\frac{Le}{L_0} = 10^{-kD} \qquad (6\text{-}27)$$

Removable BOD is defined as the maximum fraction of applied BOD removal at a specified hydraulic loading range. The constants k and L have been estimated at 0·175 and 0·90 for low rate filters treating settling sewage at a hydraulic loading of 2–6 Mgal/acre/day. For high rate filters (hydraulic loading 20 Mgal/acre/day) k and L were found to be 0·1505 and 0·784 respectively.

The Velz formula indicates that for a given bed depth the removal efficiency is constant to a limiting load, beyond which no further BOD is removed. The limiting load has been specified as 1·0 lb BOD/ft^2/day at 30°C.

The limiting load is related to the storage capacity of the bed which is in turn related to the surface area per unit volume. Equation (6-27) applies to a range of specified hydraulic loading. Incorporating hydraulic loading in Equation (6-27) results in a form of Equation (6-25).

A modification of Velz's formulation has been developed by Stack (1957). A coefficient of biosorption is employed which is defined as

$$f = 1 - 10^{-k} \qquad (6\text{-}27\text{a})$$

in which f is the fractional removal per unit depth. The limiting load which will saturate a unit depth with BOD is designated as S. When a load, S, is applied to a unit depth the maximum removal is fS.

The number of depth units X_D which will have removed the maximum assimilable load of BOD is defined:

$$X_D = 1 + \frac{L + S}{fS} \qquad \text{where } L \geq S \qquad (6\text{-}27\text{b})$$

at low loadings when $L < S$, $X_D = 0$ and the filter performance may be expressed by the equation:

$$R = fL[1 + (1 - f) + (1 - f)^2 \ldots (1 - f)^{D-1}] \quad (6\text{-}27c)$$

or

$$R = fLP$$

in which R is the total removal of removable BOD.

As D becomes infinitely large P approaches $1/f$. As the load to the filter increases and part of the depth becomes saturated with BOD, that portion of the filter depth will remove its maximum assimilation capacity is $X_D fS$. The remaining portion of the filter with a depth, $D - X_D$, will remove BOD according to Equation (6-27c) and the total removal may be expressed:

$$R = XfS + f(L - XfS)\{1 + (1 - f) + (1 - f)^2 \ldots (1 - f)^{D-X-1}\} \quad (6\text{-}27d)$$

As the loading to the filter is increased, successive portions of the bed from the top down attain their maximum assimilative capacity.

The filter will be saturated when $X_D = D$ and the BOD loading which will saturate the filter is:

$$L_s = S + (D - 1)fS$$

The BOD removal is constant at DfS. The percentage removal at this maximum level of loading is:

$$\frac{R_s}{L_s} = \frac{DfS}{S + (D + 1)fS} = \frac{Df}{1 + (D - 1)f} \quad (6\text{-}27e)$$

Stack's relationships would also have to be modified for variable hydraulic loading.

Filter Design with Recirculation

Effluent recirculation tends to increase the overall efficiency of BOD removal. Recirculation will also more effectively distribute the loading on the filter and tend to smooth out high loading on the filter and high loading variations.

Recirculation can be considered in the application of Equation (6-25) or Fig. 6-16 by employing the BOD after dilution with the recirculated flow as that applied to the filter. The BOD removal through the filter is therefore based on this applied BOD rather than that of the settled sewage. The applied BOD is calculated from the relationship:

$$lo = \frac{la + Nle}{1 + N} \qquad (6\text{-}28)$$

in which lo = applied BOD after dilution with the recirculated flow
la = BOD of settled sewage
N = recirculation ratio

Example 6-6. Design a filter to obtain 85 per cent BOD removal from an initial BOD of 200 p.p.m, using a depth of 4 ft and a hydraulic loading of 20 Mgal/acre/day:

$$\frac{D^{0.67}}{Q^{0.50}} = 0.56$$

From Fig. 6-16,

$$\frac{le}{lo} = .416 \quad \text{for} \quad \frac{D^{0.67}}{Q^{0.50}} = 0.56$$

$$le = 0.15 \cdot 200 = 30 \text{ p.p.m.}$$

$$la = 200 \text{ p.p.m.}$$

$$lo = \frac{30}{0.416} = 72 \text{ p.p.m.}$$

Therefore, from Equation (6-28)

$$72 = \frac{200 + 30N}{1 + N}$$

and the required recirculation ratio, N, is $3 \cdot 0 : 1$.

AEROBIC BIOLOGICAL TREATMENT PROCESSES

It is recognized that the treatability of the recirculated flow is less than the settled sewage due to the retardant effect in the BOD removal process. This effect is not considered in the design relationships. The error introduced, however, by neglecting this effect is small for most cases.

The effect of recirculation in the Velz formula is considered as a removal through a successive depth, i.e. a recirculation ratio of one on a three-foot bed is equivalent to a single pass through a six-foot bed.

Stack (1957) has shown that when no portion of the bed is saturated the filter performance operated at a recirculation ratio of $N:1$ is:

$$\frac{(N+1)\,PfL}{1+NPf} = PfL\,\frac{N+1}{1+NPF} = R \qquad (6\text{-}29)$$

When recirculation is employed the actual application of BOD is the sum of the applied BOD in the feed and the BOD present in the recirculated material. The performance of any filter with recirculation is obtained by combining Equations (6-27c) and (6-29) (according to Stack, 1957).

$$R = XfS + f(L - XfS)\,\frac{N+1}{1+NP(D-X)^f}\,(P_D - X) \qquad (6\text{-}30)$$

Several empirical relationships have been developed for the design of trickling filters employing recirculation treating domestic sewage. One of the most common was developed by the Upper Mississippi River Basin and is specified as the Tentative 1957 Ten State Standards.

This states "Single-stage filters will be considered where ... the applied load, recirculation included, does not exceed 110 lb 5-day BOD per 1000 ft^3 (4800 lb per acre-ft) of filter media per 24 hr." Also "The recirculation system shall apply sufficient dilution to the settled sewage so that the BOD influent to the filter, recirculation included, shall not exceed three (3) times the BOD of the required settled effluent". For two-stage filters this further states "The BOD load applied to the second stage filter, recirculation included, shall not exceed two times the BOD expected in the settled effluent, when the effluent of the first stage filter shall not exceed 50 per cent." These relationships apply to a maximum hydraulic loading of 30 Mgal/acre/day.

From these criteria Rankine (1953) developed the following design equations:

Single-Stage Plans:

$$le = \frac{la}{2N + 3} \qquad (6\text{-}31)$$

Two-Stage Plants:

$$le_1 = 0{\cdot}5la \qquad (6\text{-}32)$$

$$le_2 = le_1/N_2 + 2 \qquad (6\text{-}33)$$

The application of these formulations can be illustrated by the following example:

Example 6-7. A sewage flow of 2·0 Mgal/acre/day with a 5-day BOD of 200 p.p.m. is to be treated on a trickling filter to obtain a removal of 85 per cent. Design the filter units (based on 1957 Tentative Stds.).

(a) Recirculation ratio:
(from Equation (6-31))

$$le = \frac{la}{2N + 3}$$

$$30 = \frac{200}{2N + 3}$$

$$N = 1{\cdot}83 \quad \text{use } 2{\cdot}0$$

(b) Total applied loading:

$$\text{lb BOD/day} = 2 \cdot 0 \ . \ 200 \cdot 8 \cdot 34 + 30 \ . \ 4 \cdot 0 \ . \ 8 \cdot 34$$

$$= 3370 + 1000$$

$$= 4370 \text{ lb/day}$$

Maximum loading = 4800 lb/acre-ft per day
Therefore filter volume = 0·91 acre-ft
For a filter depth of 4 ft, area = 0·228 acres.

(c) Hydraulic loading:

$$\text{Mgal/acre/day} = \frac{6 \text{ Mgal/day}}{0 \cdot 228 \text{ acres}} = 26 \text{ Mgal/acre/day}.$$

A design formula was developed by the National Research Council from operating data from plants serving military installations during World War II (1946):

$$E = \frac{100}{1 + 0 \cdot 0085 \sqrt{(W/VF)}} \quad (6\text{-}34)$$

where W = applied BOD loading in lb per day of settled waste

V = volume of filter media in acre-ft

F = recirculation factor.

Increased removal from recirculation is considered to be due to the multiple passes of the waste through the bed. The number of passes through the bed may be computed as $(1 + N)$. It is further assumed that the removability of BOD decreases as the number of passes is increased and a weighting factor $f(f < 1$ and close to 0·9 for filters) is employed to obtain the number of effective passes, F. F may be further derived to be:

$$F = \frac{1 + N}{1 + (1-f)N^2} \quad (6\text{-}35)$$

If $\bar{f}$ is 0.9, F reaches a maximum value at a recirculation ratio N of 8.

For two-stage filters including an intermediate clarifier Equation (6-34) becomes:

$$E_2 = \frac{100}{1 + 0.0085 \sqrt{\{W_1/V_2 F_2 (1 - E_1)^2\}}} \qquad (6\text{-}36)$$

in which E_2 = per cent efficiency from second-stage filter and accompanying clarifier

W_1 = lb of BOD in effluent of intermediate clarifier

V_2 = volume of filter media in acre-ft of second-stage filter

F_2 = recirculation factor for second stage

E_1 = efficiency of the first stage.

The overall efficiency of the two-stage plant can be computed.

$$E = E_1 + E_2 - E_1 E_2 \qquad (6\text{-}37)$$

If the intermediate clarifier is omitted a lower efficiency than that computed by the formula can be expected.

Oxygen Transfer

Oxygen is transferred to the film from the waste liquid passing through the filter. The filter slime receives additional oxygen by diffusion through laminar films of liquid remaining in the filter after the applied liquid has drained away.

The rate at which oxygen is absorbed by waste passing through a filter is related to the saturation deficit, the depth of the filter, the hydraulic loading to the filter, and physical variables characteristic of the filter (Water Poll. Res. Brit. 1956). Equation (3-9) may be re-expressed as follows:

$$\frac{dC}{dD} = K_h (C_s - C) \qquad (6\text{-}39)$$

and

$$K_h \sim 1/Q^n$$

The exponent n was shown to be 0·5 (Water Poll. Res. 1956).

The quantity of active film will depend on the depth of film through which oxygen will penetrate to maintain aerobic conditions. This film depth can be developed from the following relationships:

In the trickling filter during steady state operation the rate of oxygen transfer from the flowing liquid to the film can be approximated by:

$$M = \frac{D_L}{h} \cdot A \cdot (C_1 - C_2) \qquad (6\text{-}40)$$

where C_1 and C_2 are the oxygen concentration at film depths 1 and 2, D_L the diffusivity through the film and h the film thickness. and A the interfacial area.

The rate of oxygen utilization by the film may be expressed:

$$M = k_r \cdot w \cdot A \cdot h \quad (w = \text{specific gravity}) \qquad (6\text{-}41)$$

Under steady-state conditions the rate of oxygen consumption by the film will equal the rate of oxygen transfer to the film and

$$D_L/h \, A(C_1 - C_2) = k_r \, w \, Ah \qquad (6\text{-}42)$$

The maximum penetration of oxygen will occur when the oxygen concentration C_2 at a depth h is O and

$$h = \sqrt{\left(\frac{D_L \, C_1}{k_r \, w}\right)} \qquad (6\text{-}43)$$

Schulze (1957) has indicated that the maximum depth of the aerobic zone is 2–3 mm.

In the treatment of a solution of 5000 p.p.m. glucose the maximum rate of absorption of oxygen by the film was 0·105 mg O_2/cm^2 per hr. The effluent dissolved oxygen was 10 per cent of saturation (Wat. Pol. Res., 1956). The transfer coefficient of oxygen in the film can be computed from previously developed relationships.

Example 6-9. The average oxygen utilization rate of a filter film is 12 mg O_2/hr/cm^3 at 20°C. The dissolved oxygen content in the interfacial layer is 4·0 p.p.m. Compute the depth of active film. (D_L for oxygen at 20°C is 2×10^{-5} cm^2/sec)

$$h = \sqrt{\left(\frac{D_L\, C}{k_r\, w}\right)}$$

$$= \sqrt{\left(\frac{2 \times 10^{-5} \cdot 4 \cdot 3600}{1000 \cdot 12}\right)}$$

$$= 0\cdot0049 \text{ cm}$$

$$\text{rate} = 0\cdot0049 \times 12 = 0\cdot059 \text{ mg } O_2/\text{cm}^2 \text{ per hr}$$

Effect of Temperature

Temperature exerts an influence on all biological processes. Many investigators have shown, however, that the temperature influence on trickling filters is considerably less than that predicted by conventional formulae for biological oxidation. This deviation can be rationalized when filter operation is considered as a dynamic equilibrium between several rate equations. It is probable in summer operation that the oxygen transfer mechanism may exert a controlling influence due to increased respiration rates and decreased oxygen solubilities. In winter operation the respiration rate may control. In such a case the temperature dependency will be considerably less than that predicted from previously derived equations. Schroepfer (1952) observed a variation in BOD removal efficiency of 0·18 per cent per °F temperature change. Of significance is the fact that a sewage or waste temperature drop of 4–6°F may correspond to as much as a 52°F change in air temperature. Howland (1953) has shown that the temperature effect on filter performance may be expressed by the relationship:

$$E = E_{20}\, 1\cdot035^{(T-20)} \tag{6-38}$$

A possible mechanism to define the temperature effect on trickling filter efficiency can be illustrated by the following example:

Example 6-8. At 30°C the oxygen uptake rate of filter film is 20 mg O_2/hr per ml sludge. The average dissolved oxygen is 3·0 p.p.m.; at 20°C the dissolved oxygen is 4·0 p.p.m.; compute the temperature coefficient on the ultimate filter efficiency.

The temperature effect on the oxygen uptake rate of the filter film is

$$k_{r(T)} = k_{r(20)} \cdot 1 \cdot 08^{(T-20)} \text{ (see Chapter 2)}$$

The thickness of aerobic film may be approximated

$$h = \sqrt{\frac{D_L C}{k_r w}}$$

at 30°C

$$h = \sqrt{\frac{0 \cdot 097 \cdot 3 \cdot 0}{1000 \cdot 20}} \text{ at 30°C; } D_L = 0 \cdot 097 \frac{\text{cm}^2}{\text{hr}}$$

$$= 3 \cdot 83 \times 10^{-3} \text{ cm of aerobic film}$$

at 20°C

$$D_L = 0 \cdot 074 \text{ cm}^2/\text{hr}$$

$$k_{r_{20°C}} = 20/2 \cdot 15 = 9 \cdot 3 \text{ mg/hr per ml}$$

$$h = \sqrt{\frac{0 \cdot 074 \cdot 4 \cdot 0}{1000 \cdot 9 \cdot 3}}$$

$$= 5 \cdot 7 \times 10^{-3} \text{ cm of aerobic film.}$$

Total oxidation at 30°C (per cm²)

$$3 \cdot 83 \times 10^{-3} \text{ cm}^3 \cdot \frac{20 \text{ mg}}{(\text{hr})(\text{cm}^3)} = 0 \cdot 0765 \text{ mg/hr}$$

at 20°C;

$$5 \cdot 7 \times 10^{-3} \cdot 9 \cdot 3 = 0 \cdot 053 \text{ mg/hr}$$

Therefore the temperature coefficient = 1·037.

TABLE 6-6. PERFORMANCE OF HIGH RATE TRICKLING FILTERS
TREATING DOMESTIC SEWAGE*
(single stage)

Location	Filter depth ft	Hydraulic loading Mgal/acre/day	Recirculation ratio	Applied BOD p.p.m.	BOD loading lb/acre ft	% BOD reduction (settled sewage)
Fremont, Ohio	3·3	19·0	1·5	93	1780	77·5
Greak Neck, N.Y.	4·0	7·68	1·15	120	888	83·1
Okla. City, Okla.	6·0	16·3	1·0	303	3400	78·0
Richmond, Wash.	4·5	19·6	2·76	111	1900	82·7
Dothan, Ala.	3·0	15·0	2·31	150	1150	87·4
Centralia, Mo.	3·0	8·65	0·95	134	1640	89·4
Storm Lake, Io.	8·0	21·50	2·14	380	2700	84·0
Aisal, Calif.	3·2	20·8	3·06	176	2320	86·6

* Data from Rankine (1953).

TABLE 6-7. PERFORMANCE OF HIGH RATE TRICKLING FILTERS
TREATING INDUSTRIAL WASTES

Industry	Filter depth ft	Hydraulic loading Mgal/acre/day	Recirculation ratio	Applied BOD p.p.m.	BOD loading lb/yd^3	BOD reduction %	Temp.
Chemical	6·0	4·4	10	1800	0·63	94	Summer
Chemical	6·0	7·2	10	2200	1·30	65	Winter
Pharmaceutical	4·0	15–20	10–23	3110	3·34	56	Variable
Pharmaceutical	6·0	54	12	3500	1·2	66	Variable
Pharmaceutical	6·0	78	18	4100	1·6	78	Variable
Pharmaceutical	3·0	—	4	1367	7·4	53	—
Kraft mill	6·0	52	2·4	117	4·4	27	14°C
Strawboard	6·0	17	10	820	1·24	54	27°C
Distillery	8·0	25	3–5	675	1·97	60	—
Distillery	6·0	6·4	4	700	0·75	93	—
Distillery	—	30	11	685	1.10	77	—
Dairy	4·0	15	13·5	1160	1·42	92	20°C
Sewage	4·4	16	1·86	184	1·22	83·4	Variable

Filter Performance—Industrial Wastes

The single-stage biological filtration of chemical wastes (Majewski et al.) showed 92–96 per cent BOD reduction from an initial BOD of 1800 p.p.m. under summer temperature conditions. The hydraulic loading was 4·4 Mgal/acre/day with a 10·1 recycle ratio and an organic loading of 0·14 lb BOD/ft²/day. Winter operation with an applied BOD of 2200 p.p.m. and a 10 : 1 recycle ratio produced 65 per cent BOD reduction. During this operation the hydraulic load was 7·2 Mgal/acre/day and the organic loading 0·29 lb BOD/ft²/day. Nutrients in the form of ammonium nitrate and ammonium dihydrogen phosphate were added to the waste prior to filtration.

Treatment of wastes containing alcohols, ethers and leachings from powder (Dickerson, 1949) produced 61 per cent BOD reduction in first-stage filtration at a loading of 0·9–1·9 lb BOD/yd³ and 85 per cent reduction in second-stage filtration at loadings of 0·13–0·95 lb/yd³.

Removals from dairy wastes as high as 9·5 lb BOD/yd³ have been observed at loadings of 24·5 lb/yd³. To attain high BOD removal efficiencies Trebler (1949) evaluated several two-stage recirculating filters of 3½–4½ ft deep. Their data is summarized below:

Hydraulic loading Mgal/day	Raw BOD p.p.m.	Applied loading lb/yd³	Reduction
14	770	1·93	88
14	740	1·00	92
15	1310	1·00	95·3
15	1414	1·61	92·8
17	703	0·88	86·2

Distillery wastes emanating from evaporator condensate, cooling tower overflow, doubler tailings and washings have been treated on biological filters (Davidson, 1950). These wastes are low in nutrients and require supplementary feeds. Davidson (1950) demonstrated 93 per cent BOD reduction at an organic loading of 0·75 lb/yd³ and a 4 : 1 recycle. Experimental studies by Roberts and Hardwick (1951) showed 77 per cent BOD reduction at an organic loading of

1·1 lb/yd³ with an 11 : 1 recycle ratio. This fell to 52 per cent reduction when the organic loading was increased to 2·1 lb/yd³ at a 5 : 1 recycle ratio. The hydraulic loading was 30 Mgal/acre/day in these studies.

Powder plant wastes containing decomposition products of nitrocellulose, alcohol, acetone, ether, some nitro cotton, dehydration wastes, nitric oxides, nitric and sulfuric acids and soluble organics were treated by Dickerson (1951) on two-stage biological filters. Applying a hydraulic loading of 8 Mgal/acre/day with an 8 : 1 recycle ratio to the primary filter resulted in removals of 1·0–4·5 lb/yd³. Up to a maximum loading of 2·5 lb/yd³ a BOD removal efficiency of 93 per cent was attained. Greater loadings reduced the efficiency to 83–88 per cent. With a hydraulic load of 16 Mgal/acre/day and a 16 : 1 recycle 0·02–1·16 BOD/yd³ was removed. Maximum removal of 90 per cent occurred at an organic loading of 1·1 lb/yd³.

Refinery wastes contain spent caustic wastes which have a high BOD and alkalinity and a high phenolic and naphthalenic acid salt content (Degman et al., 1952). This waste has a red color and is high in pH. Wastes from fluid catalytic cracking contain methyl ethyl ketone. The optimum temperature for biological filtration was found to be 90°F. The optimum recycle ratio in the treatment of spent caustic was found to be 4·0. The presence of oil in the waste presents difficulties in the biological oxidation system. Oil is not biologically oxidized but is adsorbed by the filter film and recovered in the final tank sludge. The maximum oil content in the applied waste was found to be 100 p.p.m. Sulfide concentrations less than 10 p.p.m. appear to have little effect on the oxidation process.

Equipment

The contact surfaces of trickling filters are usually crushed rock or gravel, 1–4 in. in size. Caution should be observed in selecting material which will not disintegrate due to the nature of the waste or the weather. In order to avoid ponding and subsequent poor distribution of flow only 5 per cent of the media should be smaller than the lowest specified size. The minimum size should be 1½ in.

Recently plastic filter media containing 25 ft² of surface area per ft³ has been developed by the Dow Chemical Co. Available data indicates higher removal efficiencies per unit volume for this media.

Waste is applied to the filter surface in a spray or fine sheet of liquid. These sprays may be either fixed or movable. The fixed sprays are formed by nozzles fed from a distribution system of fixed pipes. Movable sprays are usually produced by rotating distributors. The rotary distributors consists of a special base with a stationary center or diffuser column with a rotating manifold having pipe or rectangular fabricated steel arms which are equipped with reaction-type spray nozzles. A seal is provided between the diffuser column and the rotating manifold to prevent leakage. The center line of the distributor arms are located 8–12 in. above the level of the filter media. The arms make one revolution every 15–20 sec.

The filter under drainage system consists of a series of laterals which carry the treated waste by open channel flow to main drains leading to secondary settling tanks or final disposal. The invert gradients should be 0·5–1·0 per cent with a minimum of 0·3 per cent. The velocity of flow in the filter collection channel should be maintained at 2 ft/sec to avoid solids deposition.

REFERENCES

1. ABDUL-RAHIM, S. A., HINDIN, E. and DUNSTAN, G. H., *Public Works*, **91**, 1 (1960).
2. BRYAN, E. H. and MOELLER, D. H., Proc. Third Conf. on Biological Waste Treatment, Manhattan College (1960).
3. BUSCH, A. and KALINSKE, A. A., *Biological Treatment of Sewage and Industrial Wastes*, Vol. I (Ed. by MCCABE, B. J. and ECKENFELDER, W. W.), Reinhold Pub. Corp., New York, N.Y. (1957).
4. Cone Mills, Inc., Greensboro, N.C., pilot plant report (1959).
5. DAVIDSON, A. B., *Sew. and Ind. Wastes*, **22**, 5, 654 (1950).
6. DEGMAN, J. M., MERMAN, R. G. and DEMANN, J. G., *Proc. 7th Ind. Waste Conference, Purdue Univ.* (1952).
7. DICKERSON, B. W., *Sew. and Ind. Wastes*, **21**, 4, 685 (1949).
8. DICKERSON, B. W., Proc. 6th Ind. Waste Conference, Purdue Univ., p. 30 (1951).
9. DRYDEN, F. E., BARRETT, P., KISSINGER, J. and ECKENFELDER, W. W., *Sewage and Ind. Wastes*, **28**, 2, 183 (1956).
10. ECKENFELDER, W. W. and GRICH, E. R., *Proc. 10th Ind. Waste Conf., Purdue Univ.* (1955).
11. FAIRALL, J. M., *Sew. and Ind. Wastes*, **28**, 9, 1069 (1956).
12. GELLMAN, I. and HEUKELEKIAN, H., *Sew. and Ind. Wastes*, **25**, 10, 1196 (1953).
13. GOULD, R. H., *Proc. Amer. Soc. Civil Engrs.*, Sep. 307 (Oct. 1953).
14. GRANTHAM, G. R. *Sew and Ind. Wastes*, **23**, 10, 1227 (1951).
15. GUTIERREZ, L. V. "The hydraulics and organic removal capacity of an experimental trickling filter" Thesis, Purdue Univ. (1956).

16. HERMAN, E. R. and GLOYNA, E. F., *Sewage and Ind. Wastes*, **30**, 511, 646, 693 (1958).
17. HEUKELEKIAN, H., *Sew. Wks. J.*, **17**, 3, 516 (May 1945).
18. HEUKELEKIAN, H., *Sew. Wks. J.*, **17**, 4, 743 (July 1945).
19. HEUKELEKIAN, H., *Sew. Wks. J.*, **17**, 1, 23 (Jan. 1945).
20. HEUKELEKIAN, H., *Ind. Eng. Chem.*, **41**, 1412 (1949).
21. HASELTINE, T., *Biological Treatment of Sewage and Industrial Wastes*, Vol. I (Ed. by MCCABE, B. J. and ECKENFELDER, W. W.), Reinhold Pub. Corp. New York, N.Y. (1957).
22. HOWLAND, W. E., *Sew. and Ind. Wastes*, **25**, 2, 161 (1953).
23. HOWLAND, W. E., *Proc. 12th Ind. Waste Conf.*, Purdue Univ., p. 435 (1958).
24. HOWLAND, W. E., POHLAND, F. G. and BLOODGOOD, D. E., *Proc. Third Conf. on Biological Waste Treatment*, Manhattan College (1960).
25. IMHOFF, K. and FAIR, G. M., *Sewage Treatment*, p. 303, John Wiley and Sons (1940).
26. KEEFER, C. E. and MEISEL, J., *Water and Sewage Works*, **99**, 277 (1952).
27. KOUNTZ, R. R., *Food Eng.*, **89**, 90 (1954).
28. KOUNTZ, R. R. and FORNEY, C., *Sewage and Ind. Wastes*, **31**, 7, 810 (1959).
29. LAWS, R. and BURNS, O. B., Nat'l Council for Stream Impv. Bull (1960).
30. MAJEWSKI, F., STIEN, J. and YEZZI, T., *Proc. 10th Ind. Waste Conf.*, Purdue Univ. (1955).
31. MCDERMOTT, J. H., "Influence of media surface area upon the performance of an experimental trickling filter" Thesis, Purdue Univ. (1957).
32. MCKINNEY, R. E., *Proc. Third Conf. on Biological Waste Treatment*, Manhattan College (1960).
33. National Research Council, *Sew. Wks. J.*, **18**, 417 (1946).
34. NEMEROW, N. L., and RUDOLFS, W., *Sewage and Ind. Wastes*, **24**, 1005 (1952).
35. O'CONNOR, D. J. and ECKENFELDER, W. W., *J. Water Pollution Control Fed.*, **32**, 4, 365 (1960).
36. OSWALD, W. J. and GOTAAS, H. B., *Trans. Amer. Soc. Civil Engrs.*, **122**, 73 (1957).
37. OSWALD, W. J., HEE, R. J. and GOTAAS, H. B. "Studies on Photosynthetic Oxygenation", Univ. of Calif. Inst. of Eng. Res., Series 444 **9** (1958).
38. OSWALD, W. J., *Proc. Third Conf. on Biological Waste Treatment*, Manhattan College (1960).
39. RANKINE, R. S., *Proc. Amer. Soc. Civil Engrs.*, **79**, Sep. No. 336 (Nov. 1953).
40. RICE, W. and WESTON, R. F., paper presented at Nat'l Council for Stream Impv., White Sulphur Springs, W. Va. (Aug. 1957).
41. ROBERTS, N. and HARDWICK, J. D., *Proc. 6th Ind. Waste Conf.*, Purdue Univ., p. 80 (1951).
42. RUDOLFS, W., and AMBERG, H. R., *Sew and Ind. Wastes*, **25**, 17, 191 (1953).
43. SCHROEPFER, G. J., AL-HAKIM, M. B., SEIDEL, H. F. and ZIEMKE, N. R., *Sew. and Ind. Wastes*, **24**, 6, 705 (June 1952).
44. SHULZE, K. L., *Sew. and Ind. Wastes*, **29**, 4, 458 (1957).
45. SHULZE, K. L., *Proc. Third Biological Waste Treatment Conference*, Manhattan College (1960).
46. SINKOFF, M. D., PORGES, R. and MCDERMOTT, J. H., *J. the San. Eng. Div.*, *A.S.C.E.*, **85**, SA 6, p. 51 (1959).
47. STACK, V. T., *Sew. and Ind. Wastes*, **29**, 987 (1957).
48. TREBLER, H. A. and HARDING, H. G., *Proc. 4th Ind. Waste Conf.*, Purdue Univ., 670 (1949).

49. ULLRICH, A. H. and SMITH, M. W., *Sew. and Ind. Wastes*, **23**, 10, 1248 (1951).
50. VELZ, C. J., *Sew. Wks J.*, **20**, 4, 607 (1948).
51. Water Pollution Research, 1952 Dept. of Scientific and Industrial Research, H.M. Stationery Office, London (1952).
52. Water Pollution Research, Dept. of Scientific and Industrial Research, London, Eng. (1956).
53. ZABLATSKY, H., CORNISH, M. S. and ADAMS, J. K., *Sew. and Ind. Wastes*, **31**, 12, 1281 (1959).

CHAPTER 7

ANAEROBIC BIOLOGICAL TREATMENT PROCESSES

ANAEROBIC decomposition is employed for the treatment of organic sludges and concentrated organic industrial wastes. During the process volatile organic matter is degraded through successive steps to gaseous end products. These are primarily CO_2 and CH_4.

THEORY

Organic sludges go through two basic processes during the digestion process—liquefaction and gasification. Liquefaction occurs with extracellular enzymes which hydrolyze complex carbohydrates to simple sugars, proteins to peptides and amino acids and fats to glycerol and fatty acids. The ultimate end products of the liquefaction process are primarily volatile organic acids which are produced by "acid producing" strains of bacteria. The acids produced are primarily acetic, butyric and propionic.

During gasification the end products of liquefaction are further broken down to gaseous end products. The principal components of this gaseous mixture are CO_2 and CH_4. During a well-balanced digestion process, the processes of liquefaction and gasification occur simultaneously. The degree to which the various substances present in sewage sludges and industrial wastes will be decomposed will depend on their chemical nature. Woody type material will result in approximately a 40 per cent humus-like residue. Soluble organics are almost completely decomposed. With the exception of hydrocarbons, carbonaceous material is quantitatively converted to CH_4 and CO_2 (Buswell, 1949). Free fatty acids will undergo 80–90 per cent destruction, ester fatty acids 65–85 per cent and unsaponifiable matter 0–40 per cent (Heukelekian and Mueller, 1958).

It has been found convenient to describe the digestion process in three stages:
  (a) Acid fermentation stage.
  (b) Acid regression stage.
  (c) Alkaline fermentation stage.

ANAEROBIC BIOLOGICAL TREATMENT PROCESSES 249

These stages are shown in Fig. 7-1.

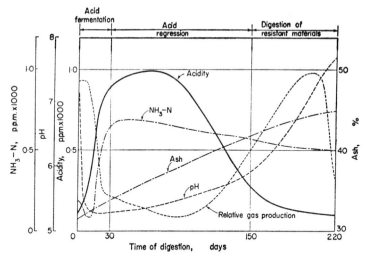

FIG. 7-1. Course of digestion of sewage solids (after Grune, 1956).

Acid Fermentation Stage

During the acid fermentation stage, carbohydrates (sugars, starches, etc.) are broken down to low molecular weight fatty acids, primarily acetic, butyric and propionic. This intensive acid production results in a drop in pH and leads to the formation of putrefactive odors. The organisms primarily responsible for this stage of the digestion process are "acid producing". Heukelekian and Mueller (1958) found that unsaturated fatty acids are hydrogenated to saturated fatty acids during acid fermentation. They also found that only 10 per cent of the grease present was degraded during this stage, leading to the conclusion that organics other than fats and grease are primarily responsible for the accumulation of volatile acids.

Acid Regression Stage

During the acid regression stage, decomposition of organic acids and soluble nitrogenous compounds occurs resulting in the formation of ammonia, amines, acid carbonates and small quantities of CO_2, N_2, CH_4 and H_2. The pH will tend to increase during this stage. By-products resulting from acid regression will include H_2S, indole, skatol and mercaptans.

Alkaline Fermentation Stage

During the alkaline fermentation stage, complete destruction of celluloses and nitrogenous compounds occur. Low molecular weight organic acids produced during the earlier stages of the process are broken down to CO_2 and CH_4. The organisms primarily responsible for the process are the spore-forming anaerobes, the methane bacteria and fat-splitting organisms. The methane bacteria are strict anaerobes, non-spore forming and require CO_2 as a hydrogen acceptor and usually organic compounds as a hydrogen donator. These organisms require an inorganic nitrogen source (NH_3). The effective pH range for methane formation is pH 6·4–7·2.

The decomposition of organic acids to CH_4 and CO_2 has been generalized by Buswell and Hatfield (1939).

$$C_nH_aO_b + (n - a/4 - b/2)\, H_2O$$
$$= (n/2 - 9/8 + b/4)\, CO_2 + (n/2 - 2/8 - b/4)\, CH_4$$

While the acid-producing bacteria seem to be resistant to high concentrations of volatile acids the methane bacteria are inhibited. Cassell and Sawyer (1959) have indicated that the methane bacteria may exist in two groups. The first group is relatively resistant to volatile acid concentration and multiplies rapidly. The second group is sensitive to volatile acid concentration and is slow to multiply. It is this second group which is primarily responsible for conversion of certain volatile acids and hence dictates the time for complete digestion. As a general rule inhibition will occur at volatile acid concentrations in excess of 2000 p.p.m. Shulze (1958) has recently found, however, that the maximum tolerable concentration will vary depending on the concentration of ammonia and other cations. The maximum desirable alkalinity concentration is approximately 2000 p.p.m. as $CaCO_3$.

Gas Production

The gas produced from the digestion of sewage sludge and similar organic mixtures is composed primarily of CO_2 and CH_4 with small quantities of NH_3, H_2S, H_2, N_2 and O_2 present. Gas from a well-digesting sludge mixture will contain 25–35 per cent CO_2 and 65–75 per cent CH_4. A gas yield of 16–18 ft^3/lb of volatile matter destroyed can be expected from digesting sewage sludge. The following table from Buswell (1939) shows gas production from various sewage constituents.

Material	CH_4 %	ft^3 of gas/lb decomposed
Fats	62–72	18–23
Scum	70–75	14–16
Grease	68	17
Crude fibre	45–50	13
Protein	73	12

The rate of gas production from a batch digestion process has been formulated in terms of the auto-catalytic equation. When the ratio of seed sludge to feed is very high only the latter phase of the auto-catalytic reaction need be considered and the gas production can be approximated by a first-order reaction.

$$\frac{dy}{dt} = K(G - y) \qquad (7\text{-}1)$$

where

G = total gas generated

y = amount of gas produced in time, t.

Simpson (1960) determined K to be 0·3/day for several digesting mixtures in which sludge was removed from active digestion and permitted to batch digest to completion. Fair and Moore (1932) found K to be 0·168/day at 95°F in digestion without mixing. Shulze (1958) determined K to be 0·14/day at 92°F. Grune et al. (1958) determined K to be 0·25 at 90°F in digesting sewage sludge.

Effect of Temperature

It has been established that two types of organisms are responsible for the digestion process—the mesophilic and the thermophilic groups. The mesophilic organisms have an effective temperature range of 29–40°C (84–104°F) and the thermophilic organisms of 50–57°C (122–135°F). It seems that digestion will proceed at approximately the same rate over a wide temperature range (35°C–60°C) providing the temperature is maintained constant. Golueke (1958) has indicated that the transition temperature range shown by Fair and Moore (1932) may not be as severe as shown, due to the action of facultative thermophilic organisms. In any event it is generally agreed that once an effective temperature range is established, slight changes in temperature can result in an upset of the methane organisms and a resultant accumulation of volatile acids.

DIGESTION DESIGN

Sludge digestion tanks are usually designed by one of three parameters.
  (a) a fixed volume of capacity per person,
  (b) a fixed volume of capacity for each pound of total solids or volatile solids fed,
  (c) a fixed period of detention.

The relationship between loading and detention can be approximately expressed by the relationship $t = 62 \cdot 4S/L$ in which S is the fraction of total or volatile solids in the sludge on a dry weight basis and L is the total or volatile solids loading in lb/ft^3 per day. t is the detention period in days.

Digestion capacity based on the population served is usually specified by state health department standards. The Ten-State Committee has indicated that for unheated digestion tanks 4–6 ft^3/capita should be provided for primary sludge and 8–12 ft^3/capita for activated sludge. When heated digestion tanks are employed 2–3 ft^3/capita is provided for primary sludge and 4–6 ft^3/capita for activated sludge.

Conventional digestion tanks are usually loaded at 0·03–0·04 lb volatile solids/ft^3/day at a detention period of 30 days. Loadings in high rate digestion systems may range from 0·1–0·2 lb/ft^3/day with detention periods of 10–15 days (Schlenz, 1955).

Full-scale high rate digestion plants have been operated at loadings of 0·18 lb vs/ft^3 per day and detention periods of 14 days. Laboratory and pilot plant studies have been operated with loadings as high as 0·42 lb vs/ft^3 per day and detention periods of 5 days. In a detailed series of digestion studies Chmielowski *et al.* (1959) found a 10-day detention to be optimum, and that the rate of digestion decreased as the detention period is reduced below this value.

Recent studies by Dunstan and Hindin (1959) indicated that volatile acid concentration increases with organic loading. Without mixing and a 33-day detention period the maximum loading was observed to be 0·11 lb vs/ft^3 per day for sewage sludge. Increasing the loading also resulted in an increase in the CO_2 content of the gas, an increase in the percentage of butyric and propionic acid and an increase in the suspended solids of the supernatant. The time required to attain steady state digestion also increases with solids loading (Shulze, 1958). The concentration of sludge solids fed to the digester was not found to be limiting up to 37 per cent solids as long as the volatile acid concentration was maintained below 2000 p.p.m. (Shulze, 1959).

Conventional Sludge Digestion

In the conventional sludge digestion process, capacity is provided for digestion, sludge storage and supernatant. Raw sludge is usually added to the digester on an intermittent pumping schedule. Supernatant is withdrawn daily and digested sludge is pumped to drying beds or disposal facilities on an intermittent schedule. Periodic mixing is usually employed to minimize local concentration build-ups, to maintain a uniform temperature and to control scum layers at the digester surface. In many plants the digestion process is accomplished in two stages, in which case mixing may be employed in the first stage. Operating data for conventional digestion are summarized in Table 7-1 and Fig. 7-2. A sectional elevation of a two-stage digestion system is shown in Fig. 7-3.

High Rate Sludge Digestion

High rate sludge digestion is accomplished by:

(a) Continuous mixing of the digester contents. This makes use of the total tank capacity for active digestion as compared to approximately one-third of the tank capacity in the conventional tank. (In the conventional tank a portion of the tank volume is used for digested sludge storage and a portion for supernatant.)

(b) Use of a high concentration feed. As was previously shown, increasing the concentration of the sludge fed to the digester will result in a higher loading (lb/ft³ per day) for the same detention period.

(c) Use of a continuous feed—while very little data are available on the advantages of a continuous feed, most investigators agree that superior digestion should result from this practice.

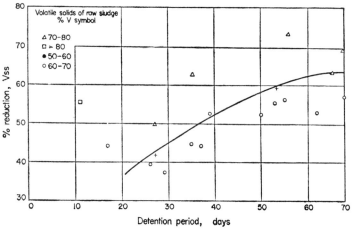

FIG. 7-2. Conventional sludge digestion performance.

The two high rate digestion systems in most common use today are the Densludge system of Dorr-Oliver Inc. and the Catalytic Reduction System of the Chicago Pump Co. In the Densludge system, raw sludge is prethickened prior to digestion. The thickened sludge is transferred to a digester containing high capacity draft tube mixers. By this procedure, no supernatant is withdrawn and the need for post-digestion thickening is eliminated. Nelson and Budd (1959) indicated that sludge prethickened to 9·1 per cent solids was successfully digested at a loading of 0·31 lb vs/ft³/day (13·7) days detention. The gas production was 16·5 ft³/lb V.M. destroyed.

The Catalytic Reduction System employs digester mixing by recirculation of gas to a diffuser manifold near the bottom of the

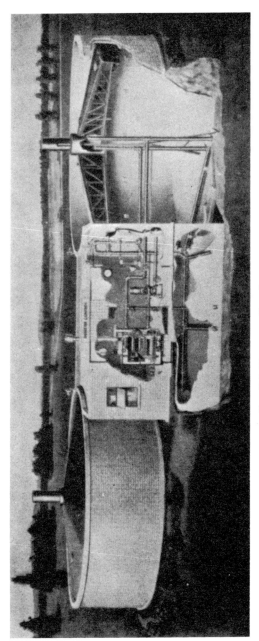

FIG. 7-3. Two-stage digestion system
(*Courtesy of Pacific Flush Tank Co.*)

FIG. 7-5 (a). High-rate digestion tanks.
(*Courtesy of Dorr-Oliver, Inc.*)

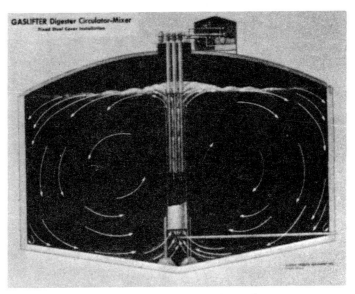

FIG. 7-5 (b). High-rate digestion tanks.
(*Courtesy of Walker Process, Inc.*)

FIG. 7-6. External sludge heat exchanger.
(*Courtesy of Walker Process Equipment, Inc.*)

ANAEROBIC BIOLOGICAL TREATMENT PROCESSES 255

digester. Sludge concentration and separation of supernatant is accomplished in secondary digestion units. Operating data for high rate digestion systems is summarized in Table 7-2 and Fig. 7-4. Some typical high rate digester layouts are shown in Fig. 7-5.

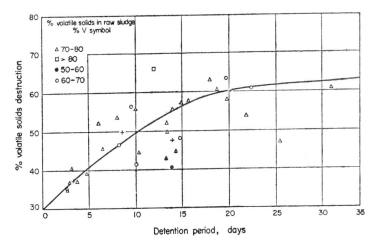

FIG. 7-4. High-rate digestion performance.

The following design procedure is taken from the A.S.C.E. recommended design procedure (1937):

Let W_d = lb dry solids added daily

V = volatile fraction

t = digestion period, days

a_t = fraction of volatiles digested in t days

Total solids remaining in

$$\text{tank at end of } t \text{ days} = W_d \cdot t - \frac{W_d t V a_t}{2} \quad (7\text{-}2a)$$

or

$$W_d t \left(1 - \frac{V a_t}{2}\right) \quad (7\text{-}2b)$$

Daily withdrawal is

$$W_d (1 - V a_{t/2}) \quad (7\text{-}3)$$

In order to have a digestion time of t days the capacity must be equal to that computed from Equation (7-2b) converted to volume by dividing by $62 \cdot 5 \, (1 - W/100)$ where W is the average moisture content of the sludge.

TABLE 7-1. CONVENTIONAL DIGESTION OPERATION

	Feed % by weight	Solids % volatile	Detention time (days)	Temp. °F	% Reduction volatile solids	Loading lb/ft³/day volatile solids	Gas prod. ft³/lb/VM destroyed
(1)	min of 5·4	73·3	33·0	heated	77·0	0·05	12·10
			33·0		77·0	0·075	11·25
			33·0		76·0	0·10	17·90
			90·0		86·0	0·075	4·54
			60·0		84·0	0·075	5·86
			30·0		83·0	0·075	9·04
			20·0		79·0	0·075	10·73
(2)	8·60	63·5	26·0	86 to 96°F	39·2	0·135	—
	7·10	63·7	35·0		44·5	0·08	—
	7·10	63·5	62·0		52·8	0·047	—
	8·10	63·4	50·0		52·3	0·067	—
	4·50	71·0	27·0		49·7	0·077	—
	8·10	67·8	39·0		52·4	0·090	—
	7·90	63·0	72·0		56·8	0·043	—
	4·90	67·7	53·0		55·2	0·040	—
	4·90	69·5	17·0		44·2	0·125	—
	9·40	67·7	55·0		56·0	0·073	—
	4·70	78·4	35·0		62·7	0·077	—
	5·50	86·4	11·0		55·3	0·265	—
	4·70	79·2	56·0		73·0	0·043	—
	3·70	64·7	37·0		44·1	0·040	—
	8·30	74·3	67·0		69·8	0·057	—
	4·80	71·9	65·0		63·9	0·033	—
	6·70	61·5	29·0		37·0	0·085	—

(1) DUNSTAN and HINDIN (1959).
(2) RANKINE, R. S. (1948).

TABLE 7-2. HIGH RATE DIGESTER OPERATION

Feed % by weight	Solids % volatile	Detention time (days)	Temp. °F	% Reduction volatile solids	Loading lb/ft³/day volatile solids	Gas prod. ft³/lb/VM destroyed
(1) 5·8	75·0	14·0	93	55·5	0·202	17·9
4·5	73·3	10·3	89	44·5	0·205	18·7
4·8	67·1	8·3	90	46·5	0·248	17·0
4·5	71·3	6·4	92	45·4	0·320	18·0
4·5	73·7	3·7	95	37·4	0·575	17·5
6·2	78·2	2·6	98	35·4	1·180	15·5
(2)	74·5	6·0	94	52·2	—	—
	74·3	8·0	—	53·5	—	—
	75·0	10·0	—	55·8	—	—
	73·7	15·0	—	57·2	—	—
	71·7	20·0	105	58·2	—	—
(3) 7·00	69·4	18·0	—	63·0	0·175	—
6·00	59·0	22·5	—	61·0	0·101	—
4·50	79·0	12·0	—	66·0	0·175	—
7·37	60·5	10·1	—	41·5	0·262	—
2·80	65·5	9·6	—	56·5	0·125	—
6·60	59·2	13·9	—	40·4	0·164	19·1
3·30	74·4	13·3	—	51·8	0·118	—
4·60	67·3	14·8	—	48·0	0·137	—
6·00	68·4	19·8	—	63·5	0·122	—
3·10	80·0	15·7	—	57·3	0·1087	—
5·10	74·5	13·3	—	42·5	0·180	—
10·20	74·1	31·0	—	61·0	0·163	—
8·20	78·0	22·0	93	53·9	0·200	14·5
8·70	78·5	18·8	94	60·8	0·240	16·5
7·60	70·9	25·5	—	47·0	0·130	—
5·00	68·6	15·9	—	68·0	0·133	—

(1) TORPEY, W. N. (1955).
(2) ROY, H. K. and SAWYER, C. N. (1955).
(3) ESTRADA, M. (1960).

Daily Additions and Periodic Withdrawals

t_0 = time for complete digestion

t_1 = time between periodic withdrawals

maximum capacity = the accumulation of sludge for t_0 days plus the accumulation of digested sludge for t_1 days

$$W_d \cdot t_0 \left(1 - \frac{V a_t}{2}\right) + W_d t_1 (1 - V_a) = \text{maximum capacity in lbs of dry solids}$$

Factors Affecting Digester Operation

In order to avoid an excessive drop in pH or an excessive build-up of volatile acids it is essential to have enough "seed" sludge for the raw sludge being fed to the digester. The seed sludge provides a balanced ratio of organisms in addition to a high buffering capacity. It has been recommended that a quantity of "seed" sludge be retained in the digestion tank equal to 10–20 times the daily input.

Sawyer and Cassell (1959) have shown that maintenance of a proper pH can be obtained by the addition of lime. The lime should be thoroughly mixed in the tank to avoid local concentration build-ups.

Mixing should be provided to avoid a build-up of a scum layer on the surface of the digesting sludge. Sawyer and Grumbling (1960) have discussed the influence of digestion time on process efficiency. The total production of gas and of methane decreases with detention time. There is also a marked decrease in grease destruction in detention periods of less than 20 days. Increasing the concentration of solids fed to the digester results in some decrease in gas quality but does not appear to influence organic solids destruction. Gas and methane yield from a 10 per cent sludge feed was 92 per cent of that produced from a 4 per cent sludge feed. Based on these factors Sawyer and Grumbling (1960) summarized the high rate digestion process as producing increased alkalinity and ammonia concentrations, decreased grease destruction and decreased volatile matter destruction.

ANAEROBIC BIOLOGICAL TREATMENT PROCESSES 259

Digester Heating

Optimum digestion of organic substances to maintain maximum gas production requires that a constant temperature in the digesting mixture be maintained. To accomplish this, internal heating coils or external heat exchangers are employed. These units are designed according to well-established principles of heat transfer.

Heat Transfer

A transfer of heat will occur when a temperature difference exists between two parts of a system. This transfer can be expressed by Fourier's Law:

$$\frac{dq}{dt} = kA \frac{dt}{dx} \qquad (7\text{-}4)$$

dq/dt = instantaneous rate of heat transfer

A = cross-sectional area of body perpendicular to direction of heat flow

dt/dx = rate change of temperature T with respect to length of heat flow path x

k = thermal conductivity of material — linear function of temperature: B.t.u./hr (ft^2) (° F/ft)

Under steady-state conditions

$$Q/t = kA \frac{\Delta T}{x} \qquad (7\text{-}5)$$

When there are several resistances in series, such as a multiple wall construction in a digestion tank Equation (7-5) can be written:

$$Q/t = \Delta T_{\text{overall}} \Big/ \frac{1}{(kA/x)_1} \times \frac{1}{(kA/x)_2} \times \frac{1}{(kA/x)_3} \qquad (7\text{-}6)$$

When heat is being transferred from a flowing fluid (as from water across a pipe wall to sludge) the primary resistance to heat transfer is considered to exist across a stagnant interfacial film. This resistance is considered as a film coefficient similar to that employed in oxygen transfer. For this case:

$$Q/t = hA \, \Delta T \qquad (7\text{-}7a)$$

in which h = film coefficient in B.t.u./(hr)(ft^2)(° F). For heat transfer through several films and a pipe wall an overall coefficient U is employed and Equation (7-7a) becomes:

$$Q/t = UA \Delta T \qquad (7\text{-}7b)$$

when

$$U = 1 \Big/ \frac{1}{h_i} + \frac{x}{k} + \frac{1}{h_0}$$

in which h_i and h_0 are the film coefficients on the sludge side and the water side respectively.

In order to design a heat exchanger it is necessary to estimate the values of the film coefficients. From dimensional analysis and the evaluation of a large mass of heat transfer data Equation (7-8) was developed for Reynolds numbers in excess of 2100 (Peters, 1954).

$$h = 0.023\, K/D \left(\frac{DV\rho}{\mu}\right)^{0.8} \left(\frac{C_p\mu}{K}\right)^{0.4} \qquad (7\text{-}8)$$

The film coefficient for the water side of the heat exchanger can be estimated from a simplification of Equation (7-8)

$$h = 150\,(1 + 0.011\, T)\, \frac{V_s^{0.8}}{D^{0.2}} \qquad (7\text{-}9)$$

in which

V = velocity of the water in ft/sec
D = inside diameter of pipe in in.
T = average temperature of flowing fluid

In the case of dense sludge flow Equation (7-8) must be modified for the non-Newtonian characteristics of the sludge.

Scale Formation

As sludge heaters are maintained in service for long periods of time, scale develops on the water side and the sludge side of the tubes which reduces the heat transfer rate. It is usual to consider the effects of this scale in terms of an additional resistance to heat transfer:

$$U = 1 \bigg/ \frac{1}{h_i} + \frac{1}{h_o} + \frac{1}{s_i} + \frac{1}{s_o}$$

(x/k for the pipe wall is neglected)

The scale coefficient for the water side of the tubes (s_i) can be selected from Table (7-3).

TABLE 7-3.

Water	125°F or less	Over 125°F
Hard water (> 15 gr/gal)	330	200
Lake and eastern mill waters	1000	500
River waters	500	330
Mississippi River and eastern rivers	330	250
Distilled water	2000	2000

An average scale coefficient for the sludge side of the tubes has been reported as 70 for temperatures of 125° F and less and 50 for temperatures greater than 125° F.

Heat Transfer Coefficients

The overall heat transfer coefficient U has been reported to vary from 10–39 B.t.u./hr ft^2 °F for internal heating coils (Greene, 1949). Scaling at Cleveland, Ohio (Polocsay, 1953) reduced the coefficient to 7·9 B.t.u./hr ft^2 °F after 1½ years' operation and to 4·4 B.t.u./hr ft^2 °F after 5½ years' operation. Overall coefficients as high as 353 B.t.u./hr ft^2 °F have been reported for new installations of forced convection heat exchanges.

Sludge Heaters

External sludge heaters presently available operate using natural convection or forced convection of hot water around the sludge piping.

The PFT heater employs natural convection of hot water around 4 in. sludge pipes. The water bath temperature is maintained at or near 145°F. The Walker Heatex unit employs water at 140°F pumped through a 6 in. pipe surrounding each 4 in. sludge pipe. The velocity of the sludge is maintained at 4 ft/sec and the hot water at 2 ft/sec.

The Dorr Spiral heat exchanger circulates sludge through one spiral while hot water is pumped counter-currently to the sludge flow. In one series of tests 224,000 B.t.u./hr of heat were transferred from a sludge flow of 75 gal/min and a water flow of 40 gal/min. The water temperature drop was 158°F to 136°F and the increase in sludge temperature 87°F to 96°F. Sludge heat exchangers are shown in Fig. 7-6.

Heat Requirements for Digestion Tanks

Heat is required during sludge digestion to
(a) raise the temperature to the desired operating level
(b) compensate for heat losses through the walls, roof and bottom of the digestion tank to the surrounding air or earth.

General heat requirements have been summarized for various ambient temperatures based on 4–5 per cent raw sludge solids in conventional digestion systems (Walker Process Equipment Co. Bull. No. 24582) (1955).

Location	Average temp. of sludge	Heat requirements B.t.u./hr/ft^3
Frigid	55°F	5–6
Temperate	50°F	3·5–4·5
Warm	45°F	3·0–4·5

More rigorous calculations can be made by modifying Equation (7-7b)

$$\text{Total heat losses} = A \left(\frac{T_i - T_o}{\frac{1}{f_i} + \frac{b}{k} + \frac{1}{f_o}} \right) \quad (7\text{-}9)$$

in which

A = area of surface, square feet
T_i = temperature of sludge in the tank °F
T_o = temperature of outside air or earth around the tank
f_i = surface coefficient of transmittance per °F to inner surface
f_o = surface coefficient of transmittance per °F from the outer surface
k = Thermal conductance of a homogeneous material 1 in. thick, B.t.u./hr/ft²/in. thickness per °F difference in temperature.
b = thickness in in.

The heat required to raise the temperature of the raw sludge is added to that computed from Equation (11) to obtain the total

TABLE 7-4. COEFFICIENTS OF THERMAL CONDUCTANCE AND SURFACE TRANSMITTANCE

Material	Conductance coefficient ("k")	Transmittance coefficient exposed to still air* ("f")	Transmittance coefficient exposed to other substances (f)
Wood (average)	1·00	1·40	—
Methane (saturated)	3·00	—	—
Air space (over ½ in.)	1·10	—	—
Air space at 50 per cent efficiency	2·20	—	—
Dry brickwork	4·00	1·40	—
Damp brickwork	5·00	—	—
Cinder concrete	5·20	—	—
Cement mortar	8·00	—	—
Stone concrete	dry wet 5·2–8·30–12·0	dry wet 1·30–1·9	—
Concrete exposed to dry earth	—	—	1·0
Steel	—	2·0	—
Concrete exposed to water or wet earth	—	—	2·0
Wrought iron exposed to sludge of 1 per cent solids	—	—	39·0

* For air motion of 15 miles/hr, multiple the given coefficient f by 3.

heat which must be supplied to maintain the proper digestion temperature.

ANAEROBIC DECOMPOSITION OF INDUSTRIAL WASTES

Anaerobic decomposition of industrial wastes has employed higher loadings and shorter detention periods than the digestion of sewage sludge. This is due primarily to the more highly soluble nature of the wastes. The results from the digestion of several wastes is summarized in Table 7-5.

TABLE 7-5. ANAEROBIC DECOMPOSITION OF INDUSTRIAL WASTES

Waste	% vol. solids	BOD		% Red.	Det'n days	Loading #/ft^3/ day	Gas ft^3/ lb	Reference
		Inf.	Eff.					
Yeast-mollasses	1·05	10,000	2,000	—	10·0	0·108	5·0	1
Butonal-grain	3·0	17,000	2,460	—	10·0	0·114	9·3	1
Yeast-mollasses	0·7	5,000	1,500	—	3·9	0·104	4·25	1
Distillery-grain	3·0	16,000	1,600	—	14·4	0·143	11·0	1
Dairy	3·76	3,300	15	—	6·0	—	—	2
Winery stillage	1·76	11,000	67	67	10·0	0·10	22·0	3

(1) BUSWELL (1949).
(2) SPAULDING (1948).
(3) PEARSON et al. (1955).

In the digestion of compressed yeast wastes 85 per cent BOD reduction was attained in less than 3 days' detention and a loading of 0·13 lb BOD/ft^3/day. The initial BOD of this waste varied from 2000–1500 p.p.m.

A detention period of 9·25 days resulted in a BOD reduction of 90 per cent from an initial BOD of 4367 in the digestion of pumpkin wastes (Hert, 1948).

A modification of the digestion process known as the anaerobic contact process has been developed for the treatment of packing-house waste (Schroepfer, 1955). This process employs a digestion unit followed by a degasifier and a settling tank from which sludge is returned to the digestion unit. Packing-house waste with a BOD

ANAEROBIC BIOLOGICAL TREATMENT PROCESSES 265

of 1500 p.p.m. was reduced in BOD by 95 per cent and suspended solids by 90 per cent at a loading level of 0·2 lb BOD/ft³ per day. The detention period in the digester was 12 hr based on the raw waste flow. The suspended solids concentration was maintained at 15,000 p.p.m.

Example 7-1. A high rate digester with a secondary digester for sludge storage is to be designed to handle primary sludge from a sewage flow of 2 Mgal/day. The initial average suspended solids in the sewage is 203 p.p.m. of which 65 per cent are volatile and 65 per cent removal can be expected from primary settling. Sludge will be pumped at 6 per cent solids from the primaries.

(a) Daily sludge loading
Suspended solids removed = 203 p.p.m. . 0·65 . 2 Mgal/day . 8·34 = 2205 lb/day.

(b) Digestion capacity—for high rate digestion at 90°F a detention period of 10 days should yield 55 per cent reduction in volatile solids. (p. 255)

Solids remaining in tank
at end of t days $= W_d t \left(1 - \dfrac{V a_t}{2}\right)$

$= 2205 \times 10 \left(1 - \dfrac{0 \cdot 65 \times 0 \cdot 55}{2}\right)$

$= 18{,}100 \text{ lb}$

Assuming the sludge mixture to have approximately the same density as water,

$$\text{volume} = \dfrac{18{,}100}{0 \cdot 06 \times 62 \cdot 4} = 4830 \text{ ft}^3$$

(c) Secondary Digester—(for sludge thickening and storage)
Assume 28 days storage at an average solids concentration of 8 per cent and 1 ft³ of sludge stored at 8 per cent contains approximately 5 lb total dry solids.

Total solids to secondary digester = 1810 lb/day

At 8 per cent solids: $\dfrac{1810}{5} = 362 \text{ ft}^3/\text{day}$

For 28 days storage digester capacity is

$$362 \times 28 = 10{,}000 \text{ ft}^3$$

Use primary digestion tank—20 ft diameter with an average depth of 15.

Heat Requirements for Winter Operation

Assume: Outside air temperature = 0° F
Ground water temperature = 40°F
Dry earth temperature = 20°F
Sludge in tank = 90°F
Sludge to tank = 50°F
Air velocity = 15 miles/hr

Tank concrete—Wall = 18 in.
Roof = 8 in.
Floor = 12 in.
Roof exposed to air

For tank 10 ft of height of walls are exposed to dry earth and 5 ft of depth and the tank bottom are exposed to ground water. The roof is in contact with gas on the inside and all other inside surfaces are in contact with sludge.

Heat Requirements to Raise Sludge Temperature

Volume of wet sludge produced per day = 4350 gal
Weight of wet sludge produced per day = 36,000 lb
Temperature increase (90–50) = 40°F
Specific heat of sludge (assume same as water)

$$36{,}000 \text{ lb/day} \times 1 \cdot 0 \frac{\text{B.t.u.}}{\text{lb °F}} \times 40°F = 1 \cdot 44 \text{ million B.t.u./day}$$

Heat Losses from Tank

Roof—Area = 314 ft^2

Using Equation (7-9)

$$\text{B.t.u./hr} = 314 \left(\frac{90° - 0°}{\frac{1}{1 \cdot 3} + \frac{8}{8 \cdot 3} + \frac{1}{3 \times 1 \cdot 3}} \right) = 14{,}200$$

Walls in dry earth

$$\text{Area} = 20 \cdot \pi \cdot 10 = 629 \text{ ft}^2$$

$$\text{B.t.u./hr} = 629 \left(\frac{90 - 20}{\frac{1}{2 \cdot 0} + \frac{18}{8 \cdot 3} + \frac{1}{1}} \right) = 12{,}100$$

Walls in ground water

$$\text{Area} = 5 \cdot \pi \cdot 20 = 314 \text{ ft}^2$$

$$\text{B.t.u./hr} = 314 \left(\frac{90 - 40}{\frac{1}{2 \cdot 0} + \frac{18}{8 \cdot 3} + \frac{1}{2}} \right) = 5{,}000$$

Floor in ground water

$$\text{Area} = 314 \text{ ft}^2$$

$$\text{B.t.u./hr} = 314 \left(\frac{90 - 40}{\frac{1}{2} + \frac{12}{8 \cdot 3} + \frac{1}{2}} \right) = 6430$$

Total heat losses from tank:

	B.t.u./hr
Roof	14,200
Walls	12,100
	5,000
Floor	6,430
	37,730 B.t.u./hr
Requirements to heat sludge	60,000
TOTAL heat requirements	97,730 B.t.u./hr (app. 100,000)

Sludge Heat Exchanger

Select an external heat exchanger with 4 in. sludge piping and a 6 in. water pipe jacketing the sludge pipe. The sludge will be pumped at a rate of 180 gal/min. The heat exchanger will use lake water from the eastern part of the United States. The water temperature entering the exchanger is 180°F and leaving is 135°F. The sludge enters at 80°F (average) and leaves at 95°F. The average overall coefficient U including scaling is estimated to be 148 B.t.u./hr °F ft. The film coefficient for sludge is assumed to be 308 B.t.u./hr °F ft.

(a) Exchanger Length

$$\Delta T_1 = 180 - 80$$

$$= 100°\text{ F (temperature gradient at inlet)}$$

$$\Delta T_2 = 135 - 95$$

$$= 40°\text{F (temperature gradient at outlet)}$$

$$\Delta T_{\text{LM}} = \frac{100 - 40}{2 \cdot 3 \log \frac{100}{40}}$$

$$= 65 \cdot 5°\text{ F (log mean temperature gradient)}$$

$$H = UA\ \Delta T_{\text{LM}}$$

$$100{,}000 = 148\ (A)\ (65 \cdot 5)$$

$$A = 10 \cdot 3\ \text{ft}^2$$

$$\text{length} = \frac{10 \cdot 3}{\pi \cdot 4/12} = 10\ \text{ft}$$

REFERENCES

1. BUSWELL and HATFIELD, *Illinois State Water Survey Bull.* (1939).
2. BUSWELL, A. M., *Proc. 5th Ind. Waste Conference, Purdue University* (1949).
3. CASSELL, A. E. and SAWYER, C., *Sew. and Ind. Wastes*, **31**, 2 (1959).

4. CHMIELOWSKI, J., SIMPSON, J. R. and ISAACS, P. G. C., *Sew. and Ind. Wastes*, **31**, 11, 1237 (1959).
5. DUNSTAN, G. H. and HINDIN, E., *Water and Sewage Works*, **106**, 10, 457 (1959).
6. ESTRADA, A., *J. S.E.D.*, *Proc. Amer. Soc. Civil Engrs.*, **86**, SA 3, 111 (1960).
7. FAIR AND MOORE, *Sew. Wks. J.* (1932) Vol. 4, 429–443.
8. GOLUEKE, C. G., *Sew. Ind. Waste J.* **30**, 10, 1225 (1958).
9. GREENE, R. A., *Sew. Wks. J.* **21**, 6, 968 (1949).
10. GRUNE, W. N. *et al.*, "Treatment and Disposal of Radioactive Sludges", Report to N.S.F., Ga. Inst. of Tech., Atlanta, Ga. (1956).
11. GRUNE, W., BARTHOLOMEW, D. D. and HUDSON, C. I., *Sew. and Ind. Wastes*, **30**, 9, 1123 (1958).
12. HERT, O. W., *Proc. 4th Ind. Waste Conference, Purdue University* (1948).
13. HEUKELEKIAN, H. and MUELLER, P., *Sew. and Ind. Wastes*, **30**, 9, 1108 (1958).
14. NELSON, F. G. and BUDD, W. I., *Proc. Amer. Soc. Civil Engrs S.E.D.* (1959).
15. PALOCSAY, F. S., *Sew. and Ind. Wastes*, **25**, 1 (1953).
15a. PEARSON, E. A., FEURSTEIN, D. F. and ONODERA, B., *Proc. 10th Ind. Waste Conf.*, Purdue Univ. (1955).
15b. PETERS, M. S. *Elementary Chemical Engineering*, McGraw Hill Book Co., New York (1954).
16. PHELPS, E. B., *Public Health Engineering*, John Wiley & Sons, Inc. (1948).
17. RANKINE, R. S., *Śew. Wks. J.*, **20**, 478 (1948).
18. ROY, H. K. and SAWYER, C. N., *Sew. and Ind. Wastes*, **27**, 1356 (1955).
19. SAWYER, C. N. and GRUMBLING, J. S., *J. S.E.D.*, *Proc. Amer. Soc. Civil Engrs.*, **86**, S.A. 2 p. 49 (1960).
20. SCHLENZ, H. E., Univ. of Kansas Bull. No. 34 (1955).
21. SCHROEPFER, G. J. *et al.*, *Sew. and Ind. Wastes*, **27**, 4, 460 (1955).
22. SHULZE, K. L., *Sew. and Ind. Wastes*, **30**, 1, 28 (1958).
23. SIMPSON, J., *Waste Treatment* (Ed. by Isaac, P.), Pergamon Press, London (1960).
23a. SPAULDING, R. A., *Proc. 4th Ind. Waste Conf.*, Purdue Univ. (1948).
24. Standard Practice in Separate Sludge Digestion, progress report of Committee of the Sanitary Engineering Division on Sludge Digestion, Amer. Soc. Civil Engrs. Proc. 63, 39 (1937).
25. TORPEY, W. N., *Sew. and Ind. Wastes*, **27**, 121 (1955).
26. Walker Process Equipment Inc. Bull. 24582 (1955).

CHAPTER 8

SLUDGE HANDLING AND DISPOSAL

SLUDGE produced from primary and secondary treatment processes usually require dewatering prior to final disposal. Methods in common use today include air drying, vacuum filtration, centrifugation and mechanical separation. The design factors involved in these processes are discussed in this chapter.

SLUDGE DRYING BEDS

One of the most common methods of dewatering sludge is by drying on open or covered sand beds. Sewage sludge will dry by this method to about 65 per cent moisture content. The area required for drying is primarily determined by climatic conditions. Area requirements can be reduced by employing glass covered beds. Sand beds are usually partitioned and contain an underdrain system to remove filtrate. Dried sludge is periodically removed from the sand beds.

Theory of Drying

Dewatering of sludge on sand beds occurs by two mechanisms; filtration of water through the sand and evaporation of water from the sludge surface. Filtration is generally complete in 1–2 days and, depending on the nature of the sludge, may result in a concentration increase to 13–22 per cent solids. Moisture removal by filtration for several sludges is shown in Table 8-1. The filtration rate can be formulated in a manner similar to that employed for vacuum filtration. The filtration can usually be increased by the addition of coagulants such as alum or ferric sulfate. When exposed to air, sludge will further dry to an equilibrium moisture content. This equilibrium or final moisture content depends upon the temperature and relative humidity of the air in contact with the sludge and the nature of the water content. A high bound water (water retained in capillaries and in cell or fibre walls) will result in a high equilibrium moisture content. Activated sludge and some primary sludges have a high bound water content.

TABLE 8-1. MOISTURE REMOVAL BY FILTRATION ON SAND DRYING BEDS

Sludge	Initial solids content (%)	Solids content after filtration (%)
Activated (pulp and paper mill waste)	2·32	16·3
Primary (pulp and paper mill waste)	3·03	22·1

Evaporation of water occurs primarily by convection or air drying. This drying process occurs in three stages, namely: a constant rate stage, a falling rate stage and a subsurface drying stage. During the constant rate period, the sludge surface is completely wetted and the rate of evaporation is independent of the nature of the sludge. This rate will be approximately the same as evaporation from a free liquid surface and will depend primarily on the air temperature, air velocity and the relative humidity.

When a critical moisture content is reached, water no longer reaches the surface of the sludge as rapidly as it evaporates and a falling rate period will occur. The rate of drying during the falling rate period will be a function of the thickness of the sludge layer, the physical and chemical properties of the sludge and the atmospheric conditions. This period is followed by subsurface drying until the equilibrium moisture content is reached. The drying process for two sludges is shown in Fig. 8-1.

Sludge Drying Bed Design

Sludge drying beds usually consist of 4–6 in. of sand over 8–10 in. of graded gravel or stone. The bed is drained by the underdrains placed 8–10 ft apart. Open sludge beds filled 8–12 in. will generally dry 5 batches of sludge per year in the northern part of the United States. The drying area can be reduced by 25–50 per cent in glass covered beds. Area requirements and sludge loading for open sludge drying beds with sewage sludge have been summarized by Imhoff and Fair (1954).

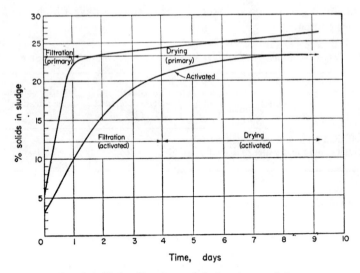

Fig. 8-1. Sludge filtration and drying characteristics.

Type of digested sludge	Area ft² per capita	Sludge loading dry solids #/ft²/year
Primary	1·0	27·5
Primary and standard trickling filter	1·6	22·0
Primary and activated sludge	3·0	15·0
Chemically precipitated sludge	2·0	22·0

Typical sludge drying data are summarized in Table 8-2. Haseltine (1951) has proposed the use of "gross bed loading" defined as the pounds of solids applied per ft² per 30 days of actual bed use and the "net bed loading" which is the product of the gross bed loading and the per cent solids in the sludge removed from the beds. This loading parameter will vary with the per cent solids applied to the sludge bed. From an analysis of 14 plants with a variation in applied sludge solids of 4–10 per cent and final sludge solids of 24·5–62·8 per cent the gross bed loading varied from 1·3–10·8.

TABLE 8-2. SLUDGE DRYING DATA

Sludge	Period	Time days	Depth applied (in.)	Solids on (%)	Solids off (%)	Notation (beds)
Digested[1]	Summer	19	9	3·8	39·3	Open
Digested[1]	Spring	20	10	4·0	42·1	Covered
Digested[1]	Spring	28	14	3·5	55·8	Covered
Digested[1]	Summer	31	10	8·75	43·3	Covered
Activated (pulp and paper mill)	Summer	4·8	12	2·3	21·9	Open
Activated (pulp and paper mill)	Summer	14	15	2·3	24·3	Open
Primary (pulp and paper mill)	Summer	9	8	3·03	26·1	Open

[1] (FLOWER et al., 1937).

VACUUM FILTRATION

Vacuum filtration is employed to separate a solid from its associated liquid by means of a porous media which retains the solid but allows the liquid to pass. Media employed for this purpose include cloth, steel mesh and tightly wound coil springs. A filter mechanism and installation are shown in Figs. 8-2 and 8-3.

The primary considerations in the filtration of a sewage or waste sludge are the filtration rate and the final moisture in the sludge cake. Filtrate solids may be an important consideration in some cases. The principal variables affecting the filtration process are the pressure drop across the sludge and the filter medium, the area of the filtering surface, the initial solids concentration and the viscosity of the filtrate. Trubnick and Mueller (1958) have also indicated that the size and shape of the solid particles and the chemical composition of the sludge and sludge liquor are primary factors affecting filtration rate. Small, irregularly shaped particles mat the filter media reducing the rate of filtrate flow.

The filtration process can be formulated for conditions of stream line flow by application of Poiseuilles and D'Arceys laws. This theory was developed by Carman (1938) and extended by Coackley (1956).

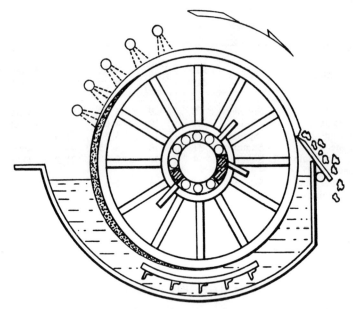

FIG. 8-2. Vacuum filter schematic (*Courtesy of Dorr-Oliver, Inc.*).

For the case of compressible filter cakes:

$$\frac{dV}{dt} = \frac{PA^2}{\mu(rcV + R_m A)} \qquad (8\text{-}1)$$

$V =$ Volume of filtrate

$t =$ Cycle time

$P =$ Pressure

$A =$ Filtration area

$\mu =$ Filtrate viscosity

$r =$ Specific resistance

$c =$ Weight of solids/unit vol. of filtrate

R_m is the initial resistance of a unit area of filtering surface. The specific resistance r is numerically equal to the pressure difference

Fig. 8-3. Coilspring vacuum filter
(*Courtesy of Komline Sanderson Corp.*)

required to produce a unit rate of filtrate flow of unit viscosity through a unit weight of cake.

Integration of Equation (8-1) yields

$$t = \frac{\mu r c}{2PA^2} \cdot V^2 + \frac{\mu R_m}{PA} \cdot V \qquad (8\text{-}2)$$

which can be rearranged to:

$$\frac{t}{V} = \frac{\mu r c}{2PA^2} \cdot V + \frac{\mu R_m}{PA} \qquad (8\text{-}3)$$

Equation (8-3) gives a linear relationship between t/V and V in which the slope $b = \mu r c / 2PA^2$ and the intercept $\mu R_m / PA$. The specific resistance r can be obtained from the relationship

$$r = \frac{2bPA^2}{\mu c} \qquad (8\text{-}4)$$

Specific resistance provides a valuable tool in the evaluation of vacuum filtration variables. While according to theory, specific resistance is independent of solids content, Coackley (1958) showed that the specific resistance decreased with decreasing solids content.

Example 8-1. Calculation of Specific Resistance.

Data:

Activated sludge suspended solids	= 2 per cent solids by weight; $C_i = 98$
Filter cake	= 17·1 per cent solids by weight; $C_f = 82·9$
Ferric chloride	= 2·5 per cent by weight
Final cake thickness	= $\frac{1}{16}$ in

$P = 10$ in. of Hg $= 352$ g/cm^2; $A = 67·8$ cm^2; $t = 20°$C

t sec	V ml	t/V
0	0	0
15	24	0·625
30	33	0·910
45	40	1·120
60	46	1·310
75	50	1·500
90	55	1·640
105	59	1·780
120	62	2·000
135	65	2·080
150	68	2·210
165	70.	2·360
180	72	2·500
195	73	2·680
210	75	2·800
225	77	2·920
240	78	3·070
285	81	3·520

$$r = \frac{2bPA^2}{\mu c}$$

$$= \frac{(2)(0·033)(352)(67·8)^2}{(0·01)(0·0226)}$$

$$= 47 \times 10^7 \text{ sec}^2/\text{g}$$

$$c = \frac{1}{\dfrac{C_i}{100 - C_i} - \dfrac{C_f}{100 - C_f}}$$

$$c = \frac{1}{\dfrac{98}{2} - \dfrac{82·9}{17·1}}$$

$$c = 0·0226 \text{ g/ml}.$$

Compressibility

Most sewage and waste applications involve compressible cakes in which the specific resistance is a function of the pressure difference across the cake. Carman (1938) proposed the relationship

$$r = r_0 P^s \tag{8-5}$$

in which s is the coefficient of compressibility. The greater the value of s the more compressible is the sludge. When $s \equiv 0$, the specific resistance is independent of pressure and the sludge is incompressible. Some typical values of specific resistance and the coefficient of compressibility are summarized in the following table:

TABLE 8-3. SLUDGE FILTRATION CHARACTERISTICS

Sludge	r (sec^2/g)*	s	Reference
Digested (Mogden, Eng.)	$1 \cdot 42 \times 10^{10}$	0·74	Coackley (1959)
Digested (Manchester, Eng.)	$1 \cdot 26 \times 10^{10}$	0·64	Coackley (1959)
Activated (fresh)	$2 \cdot 88 \times 10^{10}$	0·81	Coackley (1959)
Raw	$4 \cdot 7 \times 10^{9}$	0·54	—
Raw conditioned	$3 \cdot 1 \times 10^{7}$	1·00	Trubnick and Mueller (1958)
Digest-conditioned	$3 \cdot 3 \times 10^{7}$	1·03	Trubnick and Mueller (1958)
Digest-conditioned	$10 \cdot 5 \times 10^{7}$	1·19	Trubnick and Mueller (1958)
Digested and activated-conditioned	$14 \cdot 6 \times 10^{7}$	1·10	Trubnick and Mueller (1958)
Pulp and paper activated sludge with 2·5 per cent FeCl$_3$	$16 \cdot 5 \times 10^{7}$	0·80	—

* At 15 in. Hg.

The relationship between r and P is shown in Fig. 8-4.

Vacuum Filter Design

Equation (8-2) can be modified to predict continuous filter operation if the initial resistance of the filter system is neglected (Shepman and Cornell, 1956). (This assumption is usually valid.)

Equation (8-2) can be rewritten

$$\frac{V^2}{A^2} = \frac{2Pt_f}{\mu r c} \qquad (8\text{-}6a)$$

in which $t_f \equiv xt$ ($x \equiv$ fraction of cycle time for cake formation) Coackley (1957) has defined x as the time over which vacuum acts.

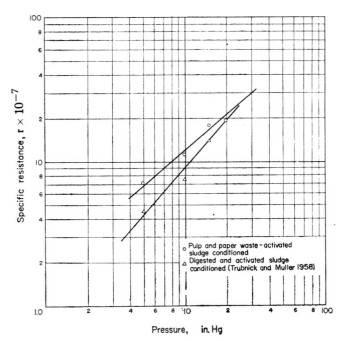

Fig. 8-4.—Relationship between filtration pressure and specific resistance.

If W = dry weight of cake = cV

$$\frac{c^2V^2}{A^2} = \frac{2Pxtc}{\mu r} = \frac{W^2}{A^2} \qquad (8\text{-}6b)$$

and

$$\frac{W}{A} = \left(\frac{2Pxtc}{\mu r}\right)^{\frac{1}{2}} \qquad (8\text{-}6c)$$

where

$$c = \frac{1}{\dfrac{C_i}{100 - C_i} - \dfrac{C_f}{100 - C_f}} \qquad (8\text{-}7)$$

The filter loading,

$$L = \frac{W}{A} \cdot \frac{60}{t} \qquad (8\text{-}8)$$

Substituting (8-8) in (8-6c)

$$L = 35 \cdot 7 \left(\frac{xcP}{\mu Rt}\right)^{\frac{1}{2}} \qquad (8\text{-}9)$$

in Equation (8-9)

$R = r \times 10^7 \text{ sec}^2/\text{g}$

$P = $ pressure, lb/in^2

$c = $ solids deposited/unit volume of filtrate, g/ml

$\mu = $ filtrate viscosity, centipoises

$t = $ cycle time, min.

Equation (8-9) can be rearranged to account for the compressibility of the sludge by substituting Equation (8-5) in Equation (8-9).

$$L = 35 \cdot 7 \left(\frac{xcP^{(1-s)}}{\mu Rt}\right)^{\frac{1}{2}} \qquad (8\text{-}10)$$

A similar equation was developed by Jones (1956)

$$L = 0 \cdot 0357 \, \frac{100 - C_f}{C_i - C_f} \left[\frac{mPC_i(100 - C_i)}{tR\mu}\right]^{\frac{1}{2}} \qquad (8\text{-}10a)$$

in which m is the fraction of time suction acts.

Effect of Operating Variables on Filtration Cycle Time

Cycle time is the time for the filter drum to complete one revolution. The total cycle time is composed of form time, drying time and dead time. Form time is the product of drum submergence and drum speed and is that portion of the cycle during which filtration occurs. Drying time is the total time of cake drying after the end of cake formation and before discharge of the cake. Dead time is that portion of the filter cycle during which the cake is being discharged and vacuum is not acting.

According to Equation (8-9) it would be expected that filter loading would vary with $t^{-\frac{1}{2}}$. Experiments on various types of sludges, however, have shown that the exponent may vary from $-0 \cdot 4$ to $-1 \cdot 0$.

This has been attributed to variations in cake permeability as additional cake is formed (Shepman and Cornell, 1956). Values obtained by these authors are summarized in the table below:

Sludge	Slope-exponent on t
Primary	0·38–0·62
Digested-primary	0·70–1·00
Elutriated-digested-primary	0·43–0·66
Digested-primary-activated	1·21
Elutriated-digested-primary-activated	0·49–0·59
Paper mill primary and activated	0·8

An interesting series of experiments on the filtration of a mixture of activated and primary sludges from a pulp and paper mill waste showed that a series of parallel lines with an exponent of 0·8 was obtained in which the filtration rate decreased with increasing activated sludge content. With 100 per cent primary sludge the filtration rate was 28 lb/ft²/hr at 15 in. Hg. When the mixture contained 40 per cent activated sludge the rate reduced to 7·5 lb/ft²/hr.

Effect of Vacuum

From Equation (8-10) the effect of vacuum on filter loading can be expressed:

$$L \sim P^{1-s/2} \sim P^n \qquad (8\text{-}11)$$

Effect of Feed Solids

An analysis of the filtration Equation (8-9) indicates that a linear relationship should exist between feed solids concentration and filter loading. As the feed solids concentration is increased, less filtrate results for each unit of cake solids deposited and the filter loading increases. Available data (Shepman and Cornell, 1956) bears out this relationship over a wide range. At very low and very high feed solids the filtration rate deviates from linearity and decreases. The slope of filtration rate–solids concentration curve varies markedly with different types of sludges.

Final Cake Moisture

Cake moisture is related to the cake thickness, the drying time, the pressure drop across the cake, the liquid viscosity, the air rate through the cake and the specific resistance. Dahlstrom and Cornell (1958) have summarized these variables in terms of a correlating factor:

$$\text{Correlating factor} = \left(\frac{\text{CFM}}{\text{ft}^2}\right)(t_d)\left(\frac{P}{d}\right)\left(\frac{1}{\mu}\right) \quad (8\text{-}12)$$

in which CFM/ft^2 is the volume of air drawn through the filter cake. The solids content of a filter cake will rise rapidly with drying time to a maximum beyond which little increase will occur. This relationship is hyperbolic in nature. Cake moisture will decrease with increasing vacuum. Beck (1955) found a cake of 80 per cent moisture at 15 in. vacuum and 78·5 per cent moisture with 18·8 in. vacuum on activated sludge conditioned with ferric chloride. The cake moisture increases with increasing cake thickness (Beck, 1955; Halff, 1952). The relationship between cake moisture content, cycle time and vacuum is shown in Fig. 8-5.

Sludge Conditioning

Because filtration rate is markedly affected by particle size and shape and by the chemical composition of the sludge and its associated liquor, sludges usually have to be conditioned by elutriation and/or addition of chemicals in order to achieve economical filtration rates. The two most common chemicals used are lime and ferric chloride. Some organic polyelectrolytes are starting to show promise for sludge conditioning.

Chemical Conditioning

Experience has shown that coagulant requirements vary widely with the type and nature of the sludge. Small particles increase the coagulant demand (Garber, 1954). Chemicals required for conditioning and subsequent filtration rate are influenced by the substances dissolved in the sludge liquor and the composition of the suspended solids (volatile and non-volatile). An increase in alkalinity or an increase in sludge volatile content increases the chemical requirement (Genter, 1946). The addition of a coagulant will result in a decrease in specific resistance and an increase in filter loading. Excessive coagulant dosages will frequently result in a reduction in filtration rate (Trubnick and Mueller, 1958). Conditioning time is

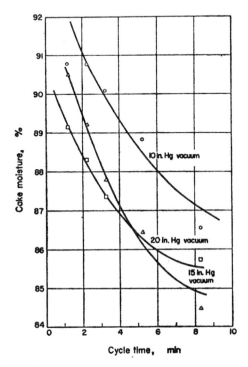

FIG. 8-5. Relationship between sludge cake moisture and filter cycle time (after Shepman and Cornell, 1958).

important to optimum filtration. Dahlstrom and Cornell (1958) reported optimum retention time of 0·25–2·0 minutes. The variation of specific resistance with ferric chloride addition is shown in Fig. 8-6 for activated sludge from a pulp and paper waste oxidation.

Elutriation

Genter (1946) showed that both the solid and liquid portions of a sewage sludge have a coagulant demand. The solids demand is related to the volatile content of the sludge and the liquid demand to the bicarbonate alkalinity of the sludge liquid. Elutriation is a solids washing process which reduces the alkalinity of the sludge liquid. The process consists of dilution with water of low alkalinity, thickening of the elutriated sludge and decantation of the elutriated

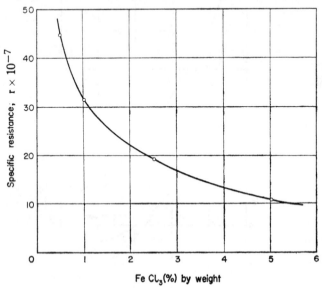

FIG. 8-6. Effect of coagulant dosage on sludge specific resistance.

liquor. Three methods of elutriation have been employed; single-stage, multistage and countercurrent. The equations for these processes have been defined by Genter (1946).

Single-stage

$$E = \frac{D + RW}{R + 1} \qquad (8\text{-}13a)$$

Multistage

$$E = \frac{D + W(R + 1)^n - 1}{(R + 1)^n} \qquad (8\text{-}13b)$$

Countercurrent

$$E = \frac{D + (R^2 + R) W}{R^2 + R + 1} \qquad (8\text{-}13c)$$

In Equations (8-13), D is the alkalinity of the sludge liquor, E the alkalinity of the elutriated sludge, W the alkalinity of the elutriating water and R the ratio of volume of elutriating water to sludge liquor. Coackley (1957) showed that the specific resistance of a digested sludge from $6 \cdot 77 \times 10^{11}$ at $11 \cdot 7$ per cent solids was reduced to $8 \cdot 3 \times 10^{10}$ at $7 \cdot 74$ per cent solids after elutriation with two volumes of water.

Laboratory Filtration Procedures

Vacuum filtration characteristics of a sludge in terms of specific resistance can be determined by the Buchner funnel test. The procedure for conducting this test is as follows:

(1) Apply vacuum to a moistened filter paper to obtain seal.
(2) Turn off vacuum and pour sample into funnel.
(3) After a suitable time to allow cake to form (5–15 sec) apply desired vacuum.
(4) Measure filtrate volume at frequent time intervals until cake cracks and vacuum drops off.
(5) Measure final cake thickness (a minimum of $\frac{1}{8}$ in. is desirable for plant design).
(6) Determine initial and final solids content in the feed sludge and final cake.

The specific resistance is calculated from the data obtained as shown in Example 8-1.

Filtration rates can also be determined by a filter leaf test in which filtration and drying are simulated on a $0 \cdot 1$ ft^2 filter surface. The test procedure is as follows:

(1) Condition approximately $2 \cdot 1$ of sludge for filtration. The sludge should be thickened to a minimum concentration of 2 per cent.
(2) Apply desired vacuum to filter leaf and immerse in sample $1\frac{1}{2}$ min (maintain sample mixed). This portion of the cycle is cake formation.
(3) Bring leaf to vertical position and dry under vacuum for 3 min (or other predetermined time). This is the cake washing and drying part of the cycle.
(4) Blow off cake for $1\frac{1}{2}$ min (this gives a total drum cycle of

6 min). To discharge the cake, the leaf is disconnected and air applied (pressure not exceeding 2 p.s.i.).

(5) Dry and weigh cake to determine percentage moisture in the cake. The filter rate in lb/ft²/hr is computed:

$$L = \frac{\text{dry weight sludge, g} \times \text{cycles/hr}}{453 \cdot 6 \times \text{test leaf area}}$$

Vacuum Filtration Performance

Results of laboratory studies using the Buchner funnel method and the leaf test for several sludges is summarized in Table 8-4. Plant performance is summarized in Table 8-5.

TABLE 8-4. LABORATORY SLUDGE FILTRATION RESULTS

Sludge	Treatment	Cake thickness mm	Cake moisture %	Feed conc. %	Vacuum in. Hg	T. °F	$r \times 10^7$ sec/g	L lb/ft²/hr
Digested	21% FeCl₃	—	83·8	2·64	—	—	10·3	—
Digested	elutriated 11·4% FeCl₃	—	78·2	2·34	—	—	65·9	—
Activated (sewage)	12·7% FeCl₃	—	90·0	1·58	—	—	33·5	—
Activated (paper mill)	—	2·3	80·2	3·28	10	74	20·6	1·24
	—	2·0	78·5	3·28	20	74	31·7	1·77
	2·5% FeCl₃	2·0	81·8	2·76	20	76	13·8	4·90
	5·0% FeCl₃	2·5	81·8	3·28	20	74	5·8	13·70
	—	1·75	79·4	2·4	20	87	50·0	1·79
	2·5% FeCl₃	2·0	83·6	2·4	20	87	12·5	9·40
	—	1·5	76·5	2·4	20	87	42·0	1·53
	2·5% FeCl₃	2·0	80·4	2·4	20	85	15·0	5·69
	5·9% FeCl₃	—	75·5	1·7	16	75	12·4	2·6
Primary (paper mill)	—	3·5	70·5	3·72	16·5	—	4·3	10·0

Cornell and Rasmussen (1958) described a design procedure where extensive pilot plant or leaf test data are available. A cycle time is selected based on an optimum cake moisture content. For this cycle

TABLE 8-5. TYPICAL PLANT PERFORMANCE-VACUUM FILTRATION

Sludge	Conc. %	Chemicals %		Cake moisture %	Filter rate lb/ft²/hr
		CaO	FeCl₃		
Primary	16·0	9·9	0	54·0	8·5
	5·5	6·0	2·2	73·0	13·0
Digested-primary	9·5	7·1	5·4	71·5	11·5
	5·6	6·0	3·0	74·0	11·0
Elutriated-digested primary	8·5	—	4·5	64·0	3·1
	9·0	—	1·5	72·5	20·0
	8·0	—	0·8	70·5	15·5
Digested-primary activated	4·5	12·0	8·0	79·0	3·0
Elutriated-digested primary activated	2·4	0	7·0	85·0	3·4
	5·0	15·0	3·0	72·5	5·0
Activated*	0·75–1·5	—	4·0–6·0	80–83	0·5–1·5
Digested†	5·6	13·5	6·0	73·3	13·9
Primary‡	6·6	8·3	3·0	73·3	14·0

* Data from IMHOFF and FAIR, *Sewage Treatment*.
† Data of SHEPMAN and CORNELL (1956).
‡ Coilspring filter.

time, a loading rate is obtained from a plot of loading vs. cycle time. If the design feed solids, vacuum or coagulant dosage differ from those employed in the loading vs. cycle time plot, the loading is proportionally adjusted using plots of loading vs. feed solids content (linear), vs. pressure (logarithmic) and vs. coagulant dosage (curvilinear). The design of a filter for a waste treatment plant is shown in Example 8-2.

Example 8-2. A waste treatment plant produces 25,000 lb of dry solids per day at a concentration of 4 per cent solids. Chemical conditioning employs 3·5 per cent ferric chloride by weight. Experimental results indicate that optimum results can be obtained using a cycle time of 3 min, 35 per cent submergence, and 20 in. vacuum, all producing an 80 per cent moisture cake. The compressibility relationship was found to be:

$$r = 0.39 \times 10^7 \, P^{0.8}$$

Design a vacuum filter for a 98 hr per week operation assuming Equation (8-10) applies

$$L = 35.7 \left(\frac{xcP}{\mu Rt}\right)^{\frac{1}{2}}$$

$$r = 0.39 \times 10^7 \, (20)^{0.8}$$
$$= 4.3 \times 10^7$$
$$\therefore R = 4.3$$

$$c = \frac{1}{\dfrac{96}{4} - \dfrac{80}{20}} = 0.05 \text{ g/ml}$$

$$L = 35.7 \left(\frac{(0.35)(0.05)(9.8)}{(1)(4.3)(3.0)}\right)^{\frac{1}{2}}$$

$$= 4.1 \text{ lb/ft}^2/\text{hr}$$

$$\frac{25{,}000 \text{ lb/day} \times 7 \text{ days/wk}}{98 \text{ hr/wk operation}} = 1780 \text{ lb/hr (solids)}$$

$$\frac{1780 \text{ lb/hr}}{4.1 \text{ lb/hr/ft}^2} = 435 \text{ ft}^2 \text{ of filter area.}$$

Mechanical Sludge Thickeners

Mechanical sludge thickeners use a vibrating screen and a rotary filter press. The sludge is introduced on the inlet end of the vibrating screen by gravity from a holding tank or by pump. The screen is equipped with combs spaced at intervals to regulate the detention of the sludge on the vibrator. The sludge is transported slowly across the vibrator while moisture is removed through filtrate lines from the bottom of the screen. The mechanism is vibrated at 1200 rev/min involving a double amplitude of 7.8 to 8.4 mm. Following the vibrator the sludge is passed through a roller press in which additional moisture is removed. The roller press may also be vibrated to facilitate additional moisture removal. Results from municipal plants in

Germany in 1954 showed a reduction in moisture content of raw sludge from 93·4 per cent to 74·5 per cent. Air drying this sludge for 24 hr further reduced the 74·5 per cent moisture content to 13·4 per cent. Extensive studies on dewatering sludge for composting at Winterthur, Switzerland, in 1956–57 involved filtering both raw and digested sludge. For raw sludge, the concentration of solids in the vibrating screen filtrate was 0·5 per cent and that in the sludge from the press was 38·6 per cent. In the digested sludge run, the feed had a solids content of 10 per cent, that in the screen filtrate was 8·8 per cent, and the concentration in the sludge from the press was 28·6 per cent solids. These Swiss studies were conducted at a feed rate of 790 gal/hr for the digested sludge and 1800 gal/hr for the fresh sludge. The Swiss studies on digested sludge showed a reduction in moisture from 88·9 per cent to 80·6 per cent through the vibrating screen and a further reduction in moisture content to 75·8 per cent through the filter press.

The unit uses no chemical conditioning or additives and operates without heating the sludge. The usual power requirement is 3 to 4 kW. The unit is operated intermittently, frequency depending on the volume and nature of sludge.

A principal objection to this unit is the high solids content in the filtrate which requires further treatment.

REFERENCES

1. BECK, A. J. et al., *Sew. and Ind. Wastes*, **27**, 6, 689 (1955).
2. CARMAN, P. C., *J. Soc. Chem. Ind.* (Brit.) **52**, 280T (1938).
3. COACKLEY, P. and JONES, B. R. S., *Sew. and Ind. Wastes*, **28**, 8, 963 (1956).
4. COACKLEY, P., *Biological Treatment of Sewage and Industrial Wastes*, Vol. II (Ed. by MCCABE, B. J. and ECKENFELDER, W. W.) Reinhold Pub. Corp., New York, N.Y. (1958).
5. COACKLEY, P., *Trade Wastes* (Ed. by ISAAC, P.) Pergamon Press, London (1960).
6. CORNELL, C. F. and RASMUSSEN, A., Report on Sludge Filtration to the West Virginia Pulp and Paper Co., Covington, Va. (1958).
7. COULSON, J. M. and RICHARDSON, J. F., *Chemical Engineering*, p. 414, McGraw Hill Book Co., New York, N.Y. (1956).
8. DAHLSTROM, D. A. and CORNELL, C. F., *Biological Treatment of Sewage and Industrial Wastes*, Vol. II (Ed. by MCCABE, B. J. and ECKENFELDER, W. W.) Reinhold Pub. Corp., New York, N.Y. (1958).
9. FLOWER, BUDD and HAUCK, *Sew. Wks. J.*, **10**, 714 (1937).
10. GARBER, W. F., *Sew. and Ind. Wastes*, **26**, 1202 (1954).
11. GENTER, A. L., *Trans. Amer. Soc. Civil Engrs*, **111**, 635 (1946).

12. GRACE, H. P., *Chem. Eng. Prog.*, **49**, 303, 367, 427 (1953).
13. HALFF, A. H., *Sew. and Ind. Wastes*, **24**, 8, 962 (1952).
14. HASELTINE, T. R., *Sew. and Ind. Wastes*, **23**, 9, 1065 (1951).
15. IMHOFF and FAIR, *Sewage Treatment*, John Wiley and Sons, New York, N.Y. (1956).
16. JONES, B. R. S., *Sew. and Ind. Wastes*, **28**, 9, 1103 (1956).
17. SHEPMAN, B. A. and CORNELL, C. F., *Sew. and Ind. Wastes*, **28**, 12, 1443 (1956).
18. TRUBNICK, E. H. and MUELLER, P. K., *Biological Treatment of Sewage and Industrial Wastes*, Vol. II, (Ed. by MCCABE, B. J. and ECKENFELDER, W. W.) Reinhold Pub. Corp., New York, N.Y. (1958).
19. YOUNG, C. H., *Sew. Wks. J.*, **8**, 331 (1935).

AUTHOR INDEX

ABDUL-RAHIM, S. A., 230
ABSON, J. W., 73
ADENEY, W. E., 88
AMBERG, H. R., 208
ANDERSON, N. E., 171

BARNHART, E. L., 92
BECK, A. J., 281
BECKER, H. G., 88
BEHN, V. C., 171
BLOODGOOD, D. E., 162
BOGAN, R. H., 34
BRYAN, E. H., 227
BUDD, W. I., 179, 254
BURNS, O. B., 114, 207
BUSCH, A. W., 28, 214
BUSWELL, A., 22, 250
BUSWELL, A. M., 248, 251, 264

CAMP, T. R., 129, 154, 156, 158, 162, 169, 170
CAPPOCK, P. D., 88
CARMAN, P. C., 273, 276
CARPINI, R. E., 96
CARVER, C. E., 87, 88
CASSELL, A. E., 250, 258
CHMIELOWSKI, J., 253
CHURCHILL, M. A., 135
CLEVENGER, G. H., 173
COACKLEY, P., 273, 275, 277, 284
COE, H. S., 173
CONE MILLS, INC., 212, 213
COOPER, C. M., 95
CORNELL, C. F., 277, 278, 280, 281, 282, 285, 286
COULSON, J. M., 171
CULLEN, E. J., 94
CUMMINGS, E. W., 168

DAHLSTROM, D. A., 281, 282
DALLAS, J. L., 163
DANCKWERTZ, P. V., 82
DAVIDSON, A. B., 243
DAVIDSON, J. F., 94
DAWSON, P. S. S., 39, 69
DEGMAN, J. M., 244
DICKERSON, B. W., 243, 244

DOBBINS, W. E., 82, 162
DORR-OLIVER, 163, 274
DOWNING, A. L., 90, 93, 98, 117, 118
DREIER, D. E., 102, 103, 104
DRESSER, 129
DRYDEN, F. E., 207, 208
DUNSTAN, G. H., 253, 256

ECKENFELDER, W. W., 14, 17, 24, 38, 40, 54, 58, 87, 92, 157, 194, 208, 212, 213
ENGLEBRECHT, R. S., 35
ESTRADA, A., 257

FAIR, G. M., 25, 26, 38, 163, 190, 251, 252, 271, 286
FAIRALL, J. M., 231
FITCH, E. B., 152, 169, 171, 173
FLOWER, 273
FROHLICH, R., 180

GADEN, E., 39, 43, 48, 94, 101
GADEN, E. L., 49, 96
GAMESON, A. H., 76, 77
GAMESON, A. L. H., 25
GARBER, W. F., 281
GARRETT, T. M., 22
GEHM, H., 44, 60
GELLMAN, I., 33, 54, 220
GENTER, A. L., 282, 283
GEYER, J. C., 25, 163
GLEASON, G. W., 129
GLOYNA, E. F., 191
GOLUEKE, C. G., 252
GOTAAS, H. B., 192, 193
GOULD, R. H., 220
GRANT, S., 48
GRANTHAM, G. R., 228
GREENE, R. A., 261
GREENHALGH, R. E., 27
GRICH, E. R., 208, 212, 213
GRUMBLING, J. S., 258
GRUNE, W. N., 249, 251
GUTIERREZ, L. V., 227

HALFF, A. H., 281
HARDWICK, J. D., 243

AUTHOR INDEX

HASELTINE, T., 204, 208, 272
HASLAM, R. T., 91
HATFIELD, 250
HAUER, R., 100
HAWKES, H. A., 34
HAYS, T. T., 183, 185
HAZEN, A., 154
HELMERS, E. N., 37, 54, 55, 62, 65
HERMAN, E. R., 191
HERT, O. W., 264
HEUKELEKIAN, H., 33, 54, 55, 171, 208, 209, 220, 224, 248, 249
HIGBIE, J., 82
HINDIN, E., 253, 256
HIXON, A. N., 49, 101
HOBER, H., 46
HOBERMAN, W. L., 86
HOLROYD, A., 92
HOOVER, S. R., 15, 19, 36, 37, 39, 42, 53, 54, 60
HORWOOD, M. P., 28
HOWLAND, W. E., 224, 225, 226, 227, 240
HURWITZ, E., 48, 116, 117, 180, 183

IMHOFF, K., 190, 271, 286
INGERSOLL, A. C., 162
INGOLS, R., 72
IPPEN, H. T., 87, 88, 89

JASEWICZ, L., 35
JENKINS, S. H., 39, 69
JERIS, J. S., 33
JONES, B. R. S., 279

KALINSKE, A. A., 28, 154, 214
KAPLOVSKY, A. J., 144
KASS, E. A., 67
KATZ, W. J., 180, 183
KEEFER, C. E., 71, 229, 231
KEHR, R. W., 94
KING, H. R., 94, 98, 101, 103, 104, 106, 107
KITTRELL, F. W., 129
KOCHTITSKY, O. W., 129
KOUNTZ, R. R., 43, 61, 100, 216

LACY, J., 171
LAWS, R., 114, 207
LEWIS, W. K., 79
LYNCH, W. O., 94

MAJEWSKI, F., 243
MANCHESTER REPORT, 117

MCCABE, B. J., 24
MCDERMOTT, J. H., 226, 228
MCKEE, 129
MCKINNEY, R. E., 28, 33, 34, 35, 38, 41, 43, 55, 62, 216
MCNAMEE, P. H., 25
MEISEL, J., 71, 229, 231
MICKLEJOHN, G. T., 88
MOELLER, D. H., 227
MOHLMAN, F. W., 48
MOORE, E. E., 25
MOORE, E. W., 26, 38, 251, 252
MOORE, T. L., 54, 67
MORGAN, P. F., 95, 99, 102, 103, 104
MORTON, R. K., 86
MUELLER, P., 248, 249, 273, 277, 278, 282

NATIONAL RESEARCH COUNCIL., 230, 237
NELSON, D. J., 34
NELSON, F. G., 179, 254
NEMEROW, N. L., 209
NICHOLS, M., 43
NUSSBERGER, F. E., 174

O'CONNOR, D. J., 40, 82, 94, 130, 138, 157, 194
OLDSHUE, J., 77
OSWALD, W. J., 189, 190, 191, 192, 193, 194

PALOCSAY, F. S., 261
PASVEER, A., 46, 88
PEARSON, E. A., 264
PETERS, M. S., 260
PHELPS, E. B., 68, 131, 133
PHELPS, E. P., 163
PHILPOTT, 88
PLACAK, O. R., 54
PORGES, N., 15, 18, 35, 36, 37, 39, 42, 52, 53, 54, 60
PRAY, H. A., 180
PRITCHARD, D. W., 138

RANKINE, R. S., 230, 236, 242, 256
RASMUSSEN, A., 285
RICE, W., 212
RICHARDSON, J. F., 171
ROBERTS, E. J., 170, 178
ROBERTS, N., 243
ROBERTSON, H. B., 76, 77
ROHLICH, G. A., 68

AUTHOR INDEX

ROXBURGH, J. M., 96
ROY, H. K., 257
RUCHHOFT, C. C., 54
RUDOLFS, W., 171, 208, 209
SAWYER, C. N., 17, 22, 34, 43, 54, 68, 69, 70, 71, 94, 171, 250, 257, 258
SCHLENZ, H. E., 252
SCHROEPFER, G. J., 179, 240, 264
SCOULLER, W. D., 88
SHEPMAN, B. A., 277, 278, 280, 282, 286
SHULTZ, J. S., 96
SHULZE, K. L., 224, 225, 227, 228, 239, 250, 251, 253
SIMPSON, J., 251
SINKOFF, M. D., 226
SMITH, M. W., 212
SPAULDING, R. A., 264
STACK, V. T., 19, 232, 233, 235
STERN, A., 38
STREETER, H. W., 25, 91
STREETER, N. W., 131, 133
SYMONS, J. M., 41, 55, 62
TALMADGE, W. P., 169, 173

TAMIYA, H., 15
THERIAULT, E. J., 25
THOMAS, H. A. JR., 125, 126, 131, 163
TORPEY, W. N., 257
TREBLER, H. A., 243
TRUBNICK, E. H., 273, 277, 278, 282
TRUESDALE, G. A., 93
ULLRICH, A. H., 212
VELZ, C. J., 19, 131, 228, 232, 235
VILLFORTH, J. C., 100
VRABLIK, E. R., 180, 182, 183
WALKER, D. J., 171
WALKER PROCESS EQUIPMENT INC. 262
WARBURG, 46, 48, 49, 50, 53
WATSON, W., 88
WESTON, R. F., 14, 24, 38, 212
WHEATLAND, A. B., 25
WHITMAN, W. C., 79
WILKE, C. R., 91
WUHRMAN, K., 19, 23, 40, 47, 55, 68, 69
ZABLATSKY, H., 212, 213

SUBJECT INDEX

Acetic acid, conversion to sludge, 54
Acids, organic, conversion to sludge, 53
Acid, fermentation, anaerobic, 249
— —, regression, anaerobic, 249
Activated sludge, organisms in, 28
—, plant design, 219
— —, settling of, 167
Activated sludge process, 8, 203
— —, conventional, 204
— —, flow diagrams, 205
— —, performance, 208
Aeration, bubble, 86
— —, devices for, performance, 90
— —, diffused, 97
— —, in streams, 82
— —, kinetics, 9
— —, mechanical, 112
— —, modified, 213
— —, step, 213
— —, theory and practice, 76
Aerators, impingement, 99, 103, 107
Aero-Accelator, 214, 215
—, performance, 216
Aerobacter aerogenes, generation time, 22
Aerobic digestion, 8
Aerobic treatments, 188
Air solubility and release, 180
Alcaligenes faecalis endogenous respiration, 42
Alcohols, bio-oxidation, 33
—, conversion to sludge, 53
Algae, growth and oxygen production, 192
—, in oxidation ponds, 189
—, nitrogen requirement, 193
—, oxygen formation, 190
—, types active in ponds, 190
Alkaline fermentation, anaerobic, 250
Alkylbenzenesulphonates, effect on oxygen transfer, 93

Aloxite tubes, oxygen transfer by 103, 104
Amino-acids, conversion to sludge, 53
Ammonia still liquors, activated sludge performance, 208
Anaerobic biological processes, 248
— —, acids produced, 248
— —, gasification in, 248, 250
— —, stages in, 249
— —, temperature effect, 251
Anaerobic digestion, 8
— —, tank design, 252
Azotobacter chroococcum oxygen uptake, 43

BOD, Delaware River, 144
—, in rivers, 127
—, —, determination, 123
—, of raw and treated sewage, 13
BOD removal, and nitrogen content, 64
— —, during declining and growth phases, 23
— —, in aerated lagoons, 198
— —, in estuaries, 136
— —, in oxidation ponds, 192
— —, in trickling filters, 224
— —, summary, 28
BOD test, 1
— —, analysis of data, 4
— —, curves, 12
— —, kinetics, 11
— —, order of reaction, 11
— —, sampling for, 2
— —, seeding material, 12
Bacillus cereus in activated sludge, 28
— —, endogenous respiration, 42
Bacillus megatherium endogenous respiration, 42
Bacillus subtilis endogenous respiration, 42
Bacillus subtilis group, in biological floc, 34

SUBJECT INDEX

Bacteria, average composition, 35
—, coliform, 1
—, endogenous respiration, 42
—, generation times, 22
—, luminous, oxygen uptake, 43
—, methane, 250
—, nitrogen and protein content, 36, 37
Bacterium methylicum formaldehyde oxidation, 33
Benzaldehyde, oxidation in activated sludge, 34
Benzoic acid, oxidation in activated sludge, 34
Biochemical oxygen demand. *See* BOD
Biological filtration, 221, 222, 223
— —, performance, 231, 242
Biological floc, 34
Biological flocculation, 28
Biological oxidation, 7
 continuous processes, 27
— —, effect of pH, 69
— —, effect of temperature, 67
— —, kinetics, 28, 29, 30
— —, nutrients required, 61
— —, of alcohols, 33
— —, of benzaldehyde, 34
— —, of benzoic acid, 34
— —, of cresols, 34
— —, of formaldehyde, 33
— —, of oxalic acid, 34
— —, of polyethylene glycol derivatives, 34
— —, of pure compounds, 33
— —, principles, 14
— —, transfer and rate mechanisms, 9
Bubble size, effects on, of waste constituents, 92
Butanol, conversion to sludge, 54

COD, samples for, 2
Carbohydrates, conversion to sludge, 53
Carbon, determination in sludge organisms, 36
Carbon dioxide evolution, and oxygen uptake, 52
Carborundum plates, 103
 bubble sizes produced, 98
Catalytic reduction system, 254

Cavitators, 116
Cell material oxidation, equation of, 15
Cell synthesis, equation of, 14
Centrifugation, of sludge, 8
Chezy coefficient, 83
Chlamydomonas in ponds, 190
Chlorella in ponds, 190
Chlorella pyrenoida oxygen uptake, 43
Chlorides, in Delaware River, 144
Chromium, effect on bacterial growth, 73
Ciliates, in biological floc, 35
Clarifiers, circular, 181
—, design, 171
—, mechanisms, 177
—, settling zones, 172
—, size, 172
Clostridium butyricum, generation time, 22
Coagulant, dosage and sludge resistance, 283
Compressibility, of sludges, 276
Compression, in sludge settling, 168
Contact-stabilization process, 209, 210
— —, oxygen utilization, 211
— —, performance, 213
Copper, effect on bacterial growth, 73
Cotton kiering liquor, pH range, 71
Cresols, oxidation in activated sludge, 34

DO. *See* Dissolved Oxygen
Densludge system, 254
Deoxygenation coefficient, 147
Dextrin, conversion to sludge, 54
Diffused aeration, design, 109
— —, performance 100, 103
Digesters, anaerobic, heat requirements, 266
— —, operation, 258
Diplococcus pneumoniac, generation time, 22
Discfuser, 100, 102, 103
Dispersed growth, aeration, 208
Dissolved oxygen, determination of, 122
— —, in Delaware River, 144
— —, in estuaries, tidal effects, 138

SUBJECT INDEX

Dissolved oxygen, minimum, 145
Dorr Spiral heat exchanger, 262
Drying beds, for sludge, 270
—, —, design, 271
—, —, theory, 270

Eberthela typhoza generation time, 22
Elutriation, of sludge, 282
Epistylis in biological floc, 35
Escherichia coli, endogenous respiration, 42
—, generation time, 22
—, oxygen uptake rates, 43
Escherichia intermedium, in activated sludge, 28
Estuaries, analysis, 122
—, flushing time, 138
—, oxygen balance, 135
—, turbulent transport coefficient, 141
Ethanol, conversion to sludge, 54
Euglena gracilis, oxygen uptake, 43
— —, in ponds, 190

Fick's law, and oxygen transport, 78
Filters, oxygen transfer in, 238
—, recirculation, 233
—, —, design, 234
Filters, trickling, 7, 8, 221
—, —, construction, 244
—, —, design, 229
—, —, mechanism of BOD removal, 224
—, —, performance, 230, 242
—, —, temperature coefficient, 241
—, —, vacuum, 8
Filters, vacuum, 273, 274
—, —, calculations, 275
—, —, design, 277
—, —, laboratory tests, 284
—, —, performance, 285, 286
Filtration, of sludges 277
—, —, cycle time, 279
—, —, resistance and pressure, 278
Flagellata spp., in biological floc, 35
Flavobacterium spp. in activated sludge, 28
—, in biological floc, 34
Flocculation, in streams, 126
Flotation, 7, 179

Flotation, dissolved air system, 184
—, rise rates, 180
Flow patterns, of wastes, 5
Flushing time, in estuaries, 138
Formaldehyde, bio-oxidation, 33
Fungi, in trickling filters, 34

Gas production, anaerobic, 250
Glucose, conversion to sludge, 53
Grit chamber, 7
Grit removal, 7
Gross bed loading, 272

Heat conductivity coefficients, 263
Heat content, of activated sludge, 38
Heat exchangers, for digesters, 268
Heat transfer, in digesters, 259
Heptanoic acid, effect on bubble aeration, 92
Hydraulic shear box, oxygen transfer by, 103

Inka system, 100
Iso-propanol, conversion to sludge, 54

Kinetics, equations for, 10

Lactose, BOD removal rate, 19
—, conversion to sludge, 54
—, oxidation of, 15
Lagoons, 188
—, aerated, design, 197
—, —, BOD removal, 200
—, —, velocity gradient, 198
Light conversion efficiency, 193
Limnotus spp. in biological floc, 35

Maltose, conversion to sludge, 54
Manometric measurement of oxygen uptake, 48
Mean contact time, in filtration, 225
Methane, formation from waste constituents, 251
Methanol, conversion to sludge, 54
Methyl ethyl ketone (butanone), trickling filter treatment, 244
Microbiol sludge, 34
Micrococcus pyogenes var. *aureus*, endogenous respiration, 42
Milk, skin, BOD removal, 18

SUBJECT INDEX

Moulds, moisture content, 36
Mycobacterium phlei, endogenous respiration, 42
Mycobacterium tuberculosis, endogenous respiration, 42

Narcodia actinomorpha, in activated sludge, 28
Net bed loading, 272
Neurospora sp., oxygen uptake, 43
Nitrification, of sludge, 60
Nitrobacter spp. in biological floc, 34
Nitrogen, content in micro-organisms, 37
—, determination in sludge, 36
—, relative costs of sources, 67
—, required for algal growth, 193
—, supply to bio-oxidation systems, 66
Nitrogen cycle, in biological waste treatment, 61, 62
Nitrosomonas spp. in biological floc, 34
Nutrients, in streams, 125
—, inorganic, for biological oxidation, 61
Nutritional requirements, in bio-oxidation, 65

Opercularia spp. in biological floc, 35
Organic matter, oxidation equation, 14
Overflow rate, and suspended solids removal, 161
Oxalic acid, bio-oxidation, 34
Oxidation, total, 216
—, —, kinetics, 216
Oxygen, dissolved. *See* Dissolved Oxygen
—, properties, 76
Oxygen absorption efficiencies, 109
Oxygen balance, in estuaries, 135
— —, in streams, 131
Oxygen concentration, and bacterial respiration, 45
Oxygen demand. (*See also* BOD), 1
— —, kinetics, 11
— —, of streams, 123
Oxygen gas analysis, in uptake studies, 53
Oxygen resources, of rivers, 128

Oxygen saturation, calculation, 77
Oxygen transfer, 78
— —, effect of gas rates, 101
— —, effect of waste constituents, 104
— —, efficiency, 107
— —, in filtration, 238
Oxygen transfer coefficient, 95
— —, effect of depth, 104
Oxygen uptake, by sludges, 41
— —, calculation for aeration tank, 110
— —, rate variations, 44
Oxygen utilization, in contact stabilization process, 211
Oxygen utilization rate, 48

pH, 2
—, effects on biological oxidation, 68
Paracolobacterium aerogenoides in activated sludge, 28
Paramaecium spp., in biological floc, 35
—, oxygen uptake, 43
Phenol, aero-accelator performance on, 216
—, oxidation of, toxicity in, 73
—, removal by activated sludge, 23
Phosphorus, in micro-organisms, 36
—, relative cost of sources, 67
Planaria agilis, oxygen uptake, 43
Plastic tubes, oxygen transfer by, 103
Polarographic analysis, of oxygen uptake, 49
Pollutional characteristics of waters, 1
Polyethylene glycol derivatives, bio-oxidation, 34
Polystyrene filter media, 224
Ponds, anaerobic, 191
—, classification, 188, 191
—, facultative, 191
—, oxidation, 189
—, —, design, 190, 194
Pretreatment, 7
Protein, in micro-organisms, 36
Proteus vulgaris, endogenous respiration, 42
— —, generation time, 22
Protozoa, in biological floc, 35
—, — —, role of, 35
Pseudomonas aeruginosa, endogenous respiration, 42

SUBJECT INDEX

Pseudomonas fluorescens, endogenous respiration, 42
Pseudomonas spp. in biological floc, 34
Rate mechanisms in biological treatment, 9
Reaeration coefficient in estuaries, 143
—, of U.S. rivers, 84
Redox potential, in finding retention period, 174
Respiration, endogenous, kinetics, 42
Rhizobium japonicum and *trifolii*, generation times, 22
Rhizopoda spp., in biological floc, 35
Rivers, BOD, 127
—, oxygen resources, 129
Rotifers, in sewage treatment ponds, 189

Saccharomyces cerevisiae endogenous respiration, 42
Sampling, frequency of, 2
Sand, and grit, removal, 7
Sarcina aurantiaca endogenous respiration, 42
Scale formation, in digester heating, 260
Scenedesmus in ponds, 190
Scour, in streams, 126
Screening, 7
Sedimentation, 7, 152
—, in streams, 126
Separation, solid–liquid, 152
Seran, tubes, oxygen transfer by, 103, 104
Serratia marcescens endogenous respiration, 42
Settling, classification, 152
—, data, analysis, 158
—, discrete and flocculent, 152
—, laboratory column, 159
—, laboratory curves, 168
Settling tanks, density currents in, 176
— —, design, 165
— —, inlet and outlet devices, 176, 178
Sewage, activated sludge performance 208
—, Aero-Accelator treatment, 216
—, characteristics, 2

Sewage, contact-stabilization treatment, 213
—, domestic, activated sludge process for, 204
—, —, available nitrogen in, 62
—, —, BOD, 9, 13
—, —, contact stabilization treatment, 212
—, —, filtration of, 7
—, —, optimum pH for treating, 71
—, —, oxygen transfer in, 105
—, —, sludge, composition, 37
—, —, —, growth, 55
—, —, —, oxidation, 59
—, —, —, —, and BOD removal, 57
—, —, suspended solids and BOD, 163
—, —, trickling filter performance, 242
—, tomato-plant, 4
Sierp apparatus, 48
Simplex Hi-Cone, 116
— —, performance, 117
Skin milk solids, conversion to sludge, 53
Sludge, activated. *See* Activated sludge
—, aeration process, 46
—, composition, 37
—, conventional anaerobic digestion, 253
—, — — —, high rate, 253
—, — — —, — —, performance, 255, 257
—, — — —, performance, 254
—, dewatering, 8
—, drying of, 272
—, —, data, 273
—, filtered, effect of coagulant on resistance, 283
—, —, elutriation, 282
—, —, moisture content, 281
—, general formula, 19
—, growth of, and BOD, 16
—, —, lag phase, 17
—, handling and disposal, 270
—, nitrogen content, 63
—, production and oxidation, 53
—, settling of, 162, 174
Solar energy, amount of, 193

SUBJECT INDEX

Solids, feed, effect on vaccuum filtration, 280
Sparjer, 100, 101, 102, 103, 104, 107, 108, 109
Sphaerotilus natans in biological floc, 34
Stabilization basins, 188
— —, aerated, 194
— —, oxygen requirement, 195
Streams, aeration in, 82
—, analysis, 122
—, minimum flow, 135
—, oxygen balance, 131
—, surveys, 133
Streptomyces griseus oxygen uptake, 43
Sucrose, conversion to sludge, 53, 54
Sugars, conversion to sludge, 54
Sulfides, biological growth inhibition by, 73
Sulfite, oxidation of, 95
Sulfite liquor, pH range, 71
— —, sludge growth constant, 56
— —, sludge oxidation rate, 59
Suspended solids, 1
from BOD test, 6

Tank circulating velocity, 105
Temperature, effect on anaerobic processes, 251
—, effect on biological activity, 67
—, effect on biological filtration, 240
Thickeners, gravity, 178
—, mechanical, 287
Tides, effect on estuarine dissolved oxygen, 139
Transfer mechanisms, in biological treatment, 9
Trichoda spp., in biological floc, 35
Turbine aerators, 114
Turbulence, effect on stream BOD, 125
Turbulent transport coefficient, in estuaries, 141

Venturi diffusor, 100
Volatile solids test, 2
Volatilization, from streams, 126
Vortair aerator, 118
Vorticella spp. in biological floc, 35

Walker Heatex unit, 262

Warburg apparatus, 48
Wastes, antibiotics, pH range, 71
—, brewery, available nitrogen in, 62
—, —, characteristics, 2
—, —, quantity, 3
—, —, sludge yield from, 55, 56
—, candied fruit, pH range, 71
—, cannery, activated sludge performance, 208
—,—, Aero-Accelator performance, 216
—, —, BOD removal in lagoons, 199
—, —, characteristics, 2
—, —, contact-stabilization treatment, 212, 213
—, —, quantity, 3
—, —, respiration phase kinetics, 42
—, constituents of effects on oxygen transfer, 91
—, dairy, characteristics, 2
—, —, respiration phase kinetics, 42
—, —, sludge yield from, 54
—, —, total oxidation batch process, 217
—, —, trickling filter performance, 242, 243
—, distillery, anaerobic decomposition, 264
—, —, trickling filter performance, 242, 243
—, explosives plant, trickling fifter performance, 244
—, flow data, 2
—, industrial, anaerobic decomposition, 264
—, —, sludge growth, 56
—, laundry, characteristics, 2
—, meat, quantity, 3
—, organic, typical, 2
—, packing house, characteristics, 2
—, —, quantity, 3
—, papermaking, activated sludge treatment, 206
—, —, — —, performance, 207, 208
—, —, Aero-accelator performance, 216
—, —, BOD removal, 26, 40
—, —, — —, initial, 18
—, —, characteristics, 2
—, —, contact-stabilization treatment, 212, 213

SUBJECT INDEX

Wastes, papermaking, critical nitrogen content, 63
—, —, kraft, endogenous respiration rate, 44
—, —, —, oxygen transfer rates, 94
—, —, nitrogen supply for treatment of, 66
—, —, oxygen saturation, 77
—, —, oxygen transfer in, 105
—, —, pH range, 71
—, —, plain aeration, 214
—, —, quantity, 3
—, —, respiration phase kinetics, 42
—, —, rise rates, 182
—, —, sludge accumulation in lagoons, 199
—, —, sludge drying data, 273
—, —, sludge yield from, 54, 55, 56, 58, 59
—, —, suspended solids removal, 164
—, —, trickling filter performance, 242
—, —, zone settling, 169
—, — (*See also* Sulfite liquor)
—, petroleum (wash), quantity, 3
—, pharmaceutical, activated sludge performance, 208 .
—, —, BOD removal, 21, 30
—, —, gasification and flotation, 174
—, —, oxygen transfer in, 105
—, —, respiration phase kinetics, 42

Wastes, pharmaceutical, sludge growth constant, 56, 57
—, —, sludge oxidation rate, 59, 60
—, —, trickling filter performance, 242
—, rag rope, available nitrogen in, 62
—, — —, sludge yield from, 55
—, — — and jute, aeration data, 214
—, refinery, activated sludge performance, 208
—, —, trickling filter treatment, 244
—, slaughterhouse, pH range, 71
—, spent sulfite liquor, BOD removal, 21
—, textile, characteristics, 2
—, —. quantity, 3
—, yeast, anaerobic digestion, 264
Water, content of bacteria moulds and yeast, 35

Yeasts, moisture content, 36
—, oxygen uptake, 43
Yeast broth, spent, pH range, 71

Zinc, effect on bacterial growth, 73
Zone settling, 167
Zooglea ramigera, in activated sludge, 28
— —, in biological floc, 34